云南"兴滇英才支持计划"创新团队
云南"兴滇英才支持计划"产业创新人才　项目资助
云南省"三茶统筹"服务乡村振兴创新团队

品鉴普洱茶

·汉英双语·

王白娟　王岳飞　主编
黄雁鸿　吴广平　安珊珊　译

云南科技出版社
·昆明·

图书在版编目（CIP）数据

品鉴普洱茶/王白娟，王岳飞主编．－－昆明：云南科技出版社，2024.2
ISBN 978-7-5587-5261-2

Ⅰ.①品… Ⅱ.①王…②王… Ⅲ.①普洱茶－茶文化 Ⅳ.①TS971.21

中国国家版本馆CIP数据核字（2023）第176588号

品鉴普洱茶
PINJIAN PU'ERCHA

王白娟　王岳飞　主编

出 版 人：温　翔
责任编辑：吴　涯　张舒园
助理编辑：张翟贤　王艺桦
整体设计：长策文化
责任校对：孙玮贤
责任印制：蒋丽芬

书　　号：ISBN 978-7-5587-5261-2
印　　刷：昆明亮彩印务有限公司
开　　本：787mm×1092mm　1/16
印　　张：15.75
字　　数：419千字
版　　次：2024年2月第1版
印　　次：2024年2月第1次印刷
定　　价：98.00元

出版发行：云南科技出版社
地　　址：昆明市环城西路609号
电　　话：0871-64190978

版权所有　侵权必究

编委会名单

·主编·

王白娟　王岳飞

·副主编·

张贵景　盛玉泊　邓秀娟　刘春艳

·编委·

邵宛芳　刘晓慧　吕才有　张志刚　阮殿蓉　吴　涯　王兴华
李家华　侯　艳　王家银　陈亚平　李　峻　赵胜男　袁文侠
陈立佼　周剑云　高　峻　江冰冰　赵春雨　黄　玮　文　斌
温　艳　沈晓静　魏珍珍　邱明忠　赛屾骊　李艳芹　武天宇

·译者·

黄雁鸿　吴广平　安珊珊

主 编 简 介

王白娟，女，白族，大理剑川人，教授、博士生导师，云南农业大学茶学院院长，云南省妇联十二届执委会常委，中国茶叶学会常务理事，云南省女科技工作者协会会长，云南省有机茶产业智能工程研究中心，云南省高校智能有机茶园建设重点实验室主任，中华优秀茶教师，云南省"三八红旗手"，云南"兴滇英才支持计划"产业创新人才和"云南省茶产业人工智能与大数据应用创新团队"和"三茶统筹"服务乡村振兴创新团队带头人，"云南省勐海智慧茶产业科技特派团"团长。主要从事智慧茶产业的研究，主持中央引导地方科技发展资金项目、国家自然科学基金项目等各类科研课题40余项，在Frontiers in Plant Science等期刊发表论文近100篇，授权专利、软著60余件，出版专著10余部。带领团队研发了世界首台适合云南山地的采茶机器人，建成了世界第一个普洱熟茶智能发酵中试线。获云南省专利一等奖、云南省科技进步三等奖、华中神农奖三等奖，排名均第一。带领团队开展了一系列卓有成效的社会服务工作，为云茶产业的生态化、标准化、智慧化高质量发展作出了突出贡献。

王白娟教授

王岳飞教授

王岳飞，男，浙江大学农业与生物技术学院教授，博士生导师，国家一级评茶师，国务院学科评议组成员，浙江大学茶叶研究所所长，浙江大学茶学学科带头人，中国科协全国首席科学传播茶学专家，浙江省茶叶学会理事长，杭州中国茶都品牌促进会理事长。主要从事茶叶生物化学、天然产物健康功能与机理、茶资源综合利用等方面的教学与研究，主持国家科技支撑项目和省重大科技专项，成功研发茶终端产品30多种，发表学术论文80余篇，著作10余部，授权国家发明专利多项。荣获首届中国茶叶学会科学技术奖一等奖、中国茶叶学会青年科技奖、首届中华茶文化优秀教师、全国大学生茶艺技能大赛优秀指导教师、浙江省师德先进个人、宝钢优秀教师奖、浙江省科技奖、浙江大学教学成果一等奖、浙江大学求是创新先锋优秀共产党员标兵、浙江大学优秀班主任、2015浙江大学永平杰出教学贡献奖（100万元）、首届杰出茶人奖、首届国际杰出茶人贡献奖等荣誉。主讲国家级精品视频公开课"茶文化与茶健康"被评为"我最喜爱的中国大学视频公开课"和"中国大学素质教育精品通选课"，点击量2600多万人次。

序

我国是世界上最早发现和利用茶的国家,后传入世界各国,而今茶与咖啡、可可已经成为世界三大无酒精饮料。云南位于世界茶树起源中心,被誉为地球的茶祖母——临沧凤庆香竹等古茶树距今约3200年。云南是普洱茶的故乡,云南独有的大叶种茶树在得天独厚的地理和气候环境中孕育,加上特殊的加工工艺,让普洱茶成为云南地理标志性产品,成为中国茶叶的一颗明珠,一直以历史悠久、品质独特、保健功效显著而蜚声中外。在云南茶叶历史发展的长河中,普洱茶不断经历着岁月的洗礼,更迭着自己的风姿,展示着自己迷人的魅力,吸引着越来越多人的喜爱。王白娟教授长期从事智慧茶业方面的研究,主持国家自然科学基金、中央引导项目等多项课题。她是中华优秀茶教师、云南"兴滇英才支持计划"产业创新人才,对普洱茶的品鉴有着独到的见解。如何把普洱茶的美传递给世界上更多的爱茶人?为此我们师生共同发力,在多年品饮实践的基础上,查阅了大量资料,调研了云南各大茶区、茶山、茶企和茶叶市场,终于完成此书的撰写工作。本书文笔流畅,深入浅出,简要介绍茶文化、普洱茶基础知识、普洱茶保健功效,重点介绍了普洱茶的饮用与品鉴,是一本引导读者如何品饮普洱茶的中英文双语版科普书籍。相信通过本书,会引领更多爱茶之人了解普洱茶、走入普洱茶的美好世界。

浙江大学

前言

为面向国内外消费者介绍云南茶文化、普洱茶基础知识、普洱茶保健功效，助推云南普洱茶走向全国、走向全世界。在云南省"三茶统筹"服务乡村振兴创新团队云南"兴滇英才支持计划"创新团队和产业创新人才项目资助下，在第一主编王白娟的博士生导师、全国茶文化传播首席专家、浙江大学王岳飞教授的亲自指导下，师徒携手率领团队对2015年出版的《云南普洱茶的饮用与品鉴》一书进行了全面修订和完善，并与云南农业大学外语学院黄雁鸿老师、吴广平老师和昆明市韶山小学安珊珊老师等携手，将其译编成中英双语版读本，希望能将普洱茶知识和云南茶文化传递给全世界更多的爱茶人。

本书是一本科技成果转化书籍，也是一本面向国际推广云南普洱茶和云南特色茶文化的中英双语版科普读物，书中重点介绍了普洱茶的品鉴：

——介绍了云南的主要茶区。本书作者走访了大量的茶叶专家、茶区政府、茶人、茶农和茶叶销售者，从目前茶业界比较认可的五大普洱茶产区来介绍相关知识。

——介绍了普洱茶的品鉴。从普洱茶的品鉴要素和评审标准，来指导喜欢普洱茶的人如何品饮一款茶叶，指导消费者在琳琅满目的品牌中和纷繁复杂的各种行话、俗语中挑选到适合自己的、健康的好茶。

——介绍了有代表性的山头茶。从茶区内挑选一些近几年有代表性、较热门的16座小山头茶进行介绍。

——介绍了有代表性的普洱茶冲泡品饮细节。从细节上冲泡品饮了几款有代表性的茶叶，从外形、香气、滋味和叶底等方面，引领消费者身临其境地品鉴。

成书之际，回顾一年多来的艰辛，不禁感慨万千。衷心感谢编委成员大量的阅读相关文献书籍资料并认真编写，无数次修改和讨论；感谢浙江大学王岳飞教授给予了很多理论上的帮助；感谢云南中茶茶业有限公司、云南六大茶山茶业股份有限公司、文山隆利春茶叶销售有限公司、凤庆峡山茶业有限公司、云南德凤茶业有限公司提供的样品品鉴并参与本书的编撰工作。特别感谢凤宁号第三代传人、现任云南凤宁茶业有限公司总经理张景贵先生历时5年、行程20多万公里，把云南大小茶山跑了数遍，为文中的16座茶山介绍提供了详尽的资料，并亲自到茶山拍摄书中所用的图片。

本书是引导读者如何品饮普洱茶的科普书籍，期待能引导国内外所有爱生活、爱健康、爱茶的朋友们真正了解、饮用和品鉴普洱茶，让普洱茶走向全国、走向全世界。

作者（编者）

目录

第一章 认识茶文化 ……………… 001

第一节 什么是茶文化？\002
一、茶文化的特性 \002
二、茶文化的精神内涵 \003

第二节 茶的起源与传播 \005
一、茶的起源 \005
二、中国茶的对外传播 \008

第三节 普洱茶的历史变迁 \010
一、普洱茶名称的由来 \010
二、普洱茶的历史变迁 \010
三、普洱茶加工工艺的历史演变 \011
四、"七子饼"的由来 \012

第四节 各国茶俗 \013
一、日本茶道 \013
二、韩国茶道 \014
三、英式下午茶 \014
四、俄罗斯人的茶饮生活 \015

第二章 普洱茶基础知识 ……………… 017

第一节 茶叶的分类 \018
一、茶叶综合分类法 \018
二、"三位一体"分类法 \020

第二节 普洱茶基础知识 \022
一、什么是"普洱茶"？\022
二、普洱茶的特点 \022
三、普洱茶的分类 \023
四、普洱纯料茶与拼配茶的区别 \025

第三章　普洱茶保健功效……………………029

第一节　普洱茶降脂减肥功效 \030
一、肥胖症概述 \030
二、普洱茶对肥胖症的作用 \031

第二节　普洱茶抗疲劳与抗衰老功效 \034
一、疲劳和衰老概述 \034
二、普洱茶的抗疲劳、抗衰老作用 \035

第三节　普洱茶抗氧化及清除自由基功效 \037
一、普洱茶的抗氧化机制 \037
二、普洱茶的抗氧化功能 \038

第四章　普洱茶的饮用品鉴……………………043

第一节　普洱茶五大主要茶区 \044
一、西双版纳茶区 \044
二、普洱茶区 \048
三、临沧茶区 \049
四、保山茶区 \051
五、文山茶区 \052

第二节　普洱茶的贮存 \054
一、温度 \054
二、相对湿度 \055
三、其他贮存条件的要求 \055

第三节　普洱茶的品鉴方法 \056
一、普洱茶冲泡技艺 \056
二、普洱茶的品鉴要素 \065
三、普洱茶的审评 \081

第四节　云南名茶山和名茶品鉴 \085
一、云南名茶山介绍 \085
二、名茶品鉴 \096

参考文献：\112

/第一章/

认识茶文化

第一节 什么是茶文化？

茶文化就是人类在生产和利用茶的过程中以茶为载体的物质文化、制度文化、行为文化、心态文化的集合。从广义上来讲，茶文化包括茶的自然科学和人文科学两方面。从狭义上讲，茶文化着重于茶的人文科学，主要指茶对人精神和社会的功能。

一、茶文化的特性

茶文化作为一种文化现象，包含了作为载体的茶和人因茶而形成的各种观念形态，其必然具有自然属性和社会属性两个方面的形式和内涵。其特性主要表现为以下五个方面：

（一）社会性

随着社会的进步，饮茶文化已渗透到社会的各领域和生活的各方面。开门七件事：柴米油盐酱醋茶。即使是祭天祀地拜祖也得奉上"三茶六酒"。因此在人类发展史上无论是王公显贵还是三教九流都以茶为上品，虽然饮茶方式和品位不同，但对茶的推崇和需求却是一致的，即把饮茶当成是人类美好的物质享受与精神陶冶。而历代文人墨客、社会名流以及宗教界人士更是有雅士七事：琴棋书画诗酒茶，极大地推动了茶文化的发展。

（二）广泛性

茶文化是一种范围广泛的文化，雅俗共赏，各得其所。古老的中国传统茶文化同其他国家的历史、文化、经济及人文相结合，演变成英国茶文化、日本茶文化、韩国茶文化、俄罗斯茶文化及摩洛哥茶文化等。茶文化可以把全世界茶人联合起来，进行茶艺切磋、学术交流和经贸洽谈。

（三）民族性

据史料记载，茶文化始于中国古代巴蜀族人，后逐渐以汉族茶文化为主体并传播扩展。但每个民族因自己特有的历史文化个性使得茶文化呈现出多姿多彩的特性。蒙古族的奶茶、维吾尔族的奶茶和香茶、苗族和侗族的油茶作为日常饮食，以茶养生；白族的三道茶、苗族的三宴茶追求的是借茶喻世，寓意为人做事的哲理；傣族的竹筒香茶、傈僳族的雷响茶、回族的罐罐茶追求的是精神享受和饮茶情趣；藏族的酥油茶、德昂族的酸茶、鄂温克族的奶茶追求的是以茶为饮，并寓意示礼联谊。

（四）区域性

"千里不同风，百里不同俗。"烹茶、饮茶方法，用茶目的以及对茶叶品种需求都因地域不同而有差异。东方人推崇清饮，饮茶的基本方法是用开水直接冲泡茶叶，无需加入糖、薄荷、柠檬、牛奶、葱、姜等作料；欧美及大洋洲人钟情的是加奶、糖的红茶；而西非和北非人最爱喝的是加有薄荷或柠檬的绿茶。在中国，南方人喜欢饮绿茶，北方人崇尚花茶，福建、广东、台湾人钟情乌龙茶，西南一带推崇普洱茶，边疆民族爱喝以砖茶制作的各类调饮，等等。

（五）传承性

茶文化本身就是中国传统文化的一个重要组成部分。茶文化的社会性、广泛性、民族性、区域性决定了茶对中国文化的发展具有传承性的特点，茶文化是中华文化形成、延续和发展的重要载体。在当代特别是改革开放以后，茶文化作为民族优秀文化的组成部分，得到社会各界的认可和推崇，现代文化理念和时代新元素的融入使得茶文化价值功能更加显著、对现代化社会的作用进一步增强。其传播方式呈现大型化、现代化、社会化和国际化趋势。

二、茶文化的精神内涵

茶文化的发展历程表明，茶文化总是在满足人类社会物质生活的基础上发展成为精神生活的需要。茶文化的精神内涵主要表现为"四个结合"：

（一）物质与精神的结合

俗话说"衣食足而礼仪兴"，物质的丰富、精神生活的提高，必然促进文化的高涨。唐韦应物赞茶"洁性不可污，为饮涤尘烦"；北宋苏东坡誉茶"从来佳茗似佳人"；南宋杜耒说"寒夜客来茶当酒"；近代鲁迅认为品茶是一种"清福"；日本高僧荣西禅师称茶"上通诸天境界，下资人伦矣"。可见，茶作为一种物质其形态千姿百态，而作为一种文化又有着深邃的内涵。

(二)高雅与通俗的结合

茶文化是雅俗共赏的文化,其发展过程就是高雅和通俗共存并在统一中向前发展的过程。宫廷贵族的茶宴,僧侣士大夫的斗茶,文人墨客的品茗,以及由此派生出的有关茶的诗词、歌舞、戏曲、书画、雕塑,都是茶文化高雅性的表现。而民间的饮茶习俗非常通俗化,老少皆宜,并由此产生了茶的民间故事、传说、谚语等,这是茶文化的通俗性所在。精致高雅的茶文化植根于通俗的茶文化中,如果没有粗犷通俗的民间茶文化土壤,高雅茶文化也就失去了生存的基础。

(三)功能与审美的结合

茶在满足人类物质生活方面表现出广泛的实用性,食用、治病、解渴都需要用到茶。而"琴棋书画诗酒茶",茶又与文人雅士结缘,在精神生活需求方面,又表现出广泛的审美性。茶叶花色品种的绚丽多姿,茶文学艺术作品的五彩缤纷,茶艺、茶道、茶礼的多姿多彩,都能满足人们的审美需要。它集装饰、休闲、娱乐于一体,既是艺术的展示,又是民俗的体现。

(四)实用与娱乐的结合

随着茶综合利用的进展,茶的开发利用已扩展到多种行业。近年来,开展的多种形式的茶文化活动,如茶文化节、茶艺表演、茶文化旅游等,就是以茶文化活动为主体,满足人们在品茗、休闲、旅游的同时又能达到促进经济发展的作用,体现了实用与娱乐的结合。

总之,在茶文化中,蕴含着进步的历史观和世界观,它以健康、向上、平和的心态去鼓励人类实现社会进步的理想和目标。

第二节 茶的起源与传播

茶字经过一系列的演变,最终定形于中唐时期,有九笔画,上部为"艹",像茶的穿叶,中部"人"像树冠,下部"木"代表树干。茶是木本植物,在植物分类系统中属于被子植物门,双子叶植物纲,山茶目,山茶科,山茶属。

一、茶的起源

(一)茶的起源与原产地

中国是茶树的原产地,这一点得到世界各国的广泛认可。随着考证技术的发展和新发现,确认了中国西南地区,包括云南、贵州、四川是茶树原产地的中心。由于地质变迁及人为栽培,茶树开始由此普及全国,并逐渐传播至世界各地。茶是中华民族对世界文明的又一伟大贡献。

中国人也是最早发现、利用和栽培茶树的,神农尝百草发现茶叶的传说,距今大约有五六千年的历史。学者们普遍认为,公元3世纪之前,茶在中国就已经非常盛行了。到3000多年前的西周初期,中国人就开始栽培茶树。

世界上最粗壮的茶树——临沧凤庆香竹箐古茶树高达10.6米,树冠11.5米,树干直径1.84米,基围5.8米。其树龄据推断有3200年。在其周围还有栽培的古茶树群14000多株,这些古茶树是活化石,是人类悠久种茶、饮茶历史的有力见证。

凤庆香竹箐古茶树

全世界山茶科茶组植物共有31个种4个变种,中国有30个种和4个变种,因此中国也是当之无愧的茶树种质资源最丰富的国家。

（二）用茶起源与演变

人类对茶的利用是从药用开始，然后才发展成为食用和饮用。

"神农尝百草，日遇七十二毒，得荼而解之"，"荼"是茶的古字。经过后人长期实践，发现茶叶不仅能解毒，如果配合其他中草药，还可以治疗多种疾病。明代顾元庆在《茶谱》中写道："人饮真茶能止渴、消食、除痰、少睡、利尿、明目益思、除烦去腻。"把茶的药用功能说得非常清楚。对于中国边疆少数民族，茶的药用功能更为突出。在少数民族地区，流传着"宁可三日无粮，不可一日无茶"的谚语。这是因为，像藏族、蒙古族、维吾尔族等少数民族都是居住在高寒地区，日常主食都是牛羊等肉类食品，不易消化，而茶可以解油腻、促消化以及补充各种维生素、微量元素等。

早期的茶，除了作为药物，很大程度上还作为食物出现。吃"腌茶""茗粥""擂茶"的习俗还被部分少数民族保留了下来。

中国人利用茶的年代久远，但饮茶的出现相对要晚一点。有文献记载的是公元前59年，西汉辞赋家王褒的《僮约》中有所反映："武阳买茶，烹茶尽具"，说明在西汉时期，已经有茶叶市场和饮茶的风尚，距今已有2000多年了。饮茶习俗的形成，从西汉到三国时期，也就是公元前206年到公元260年间，在巴蜀之外，茶是供上层社会享用的一个珍稀植物，饮茶仅限于王公贵族。晋朝以后，饮茶逐步进入了中下层社会。两晋南北朝时期，公元265年到581年，上至帝王将相，中及文人士大夫、宗教徒，下到平民百姓，社会各阶层普遍饮茶，饮茶成为了中国人的一个习俗。唐朝时期，公元618—906年，饮茶之风最为盛行，被认为是茶的黄金时代，家家户户都饮茶，且流传于塞外。在此期间诞生了世界上第一部茶叶专著，即陆羽所著的《茶经》。该书总括了中国人加工茶、品饮茶、研究茶、颂扬茶的历史，系统地介绍和论述了有关茶的知识和学问，把人类饮茶从生理需要提高到文化需要。

（三）饮茶方法的演变

中国饮茶的历史经历了漫长的发展和变化时期。不同的阶段，饮茶的方法、特点都不相同。饮茶方法最初是烹煮，唐代为煎，宋代为点，到明清时期改为冲泡。

唐朝时期，嫩叶一旦被采摘下来，就通过蒸、压然后倒入模具制成饼状，烘烤至干燥。冲泡时，需将茶饼在火上炙烤，再碾压成粉末，放入水中煮沸。在一些地方，为了减少茶的苦涩味，常在茶水中加入盐，也可以在茶水沸腾前或沸腾后加入各种调料，如甜葱、生姜、橘皮、丁香、薄荷等。

到了宋代，茶的冲泡方式有所变化，紧压的茶饼先研磨成粉末，然后加水不断击拂，产生具绵密泡沫的茶汤。如果茶汤颜色呈现乳白色，茶汤表面泛起的汤花能较长时间咬住杯盏内壁不动，这样就算点泡出一杯好茶。为了评比调茶技术和茶质的优劣，宋代兴起了斗茶。

元代以前，人们饮茶时有加入各种调料与茶混煮的习惯，但到了元代逐渐被人们所摒弃。明太祖朱元璋体察民情，为减轻负担，下令贡茶改制，重散略饼，促进了散茶生产技术的发展，随之而来的是茶饮方式的简约化。因此明清使用的是更为简单的清饮方式，即以沸水直接冲泡茶叶的方法，一直沿用至今。

（四）茶叶生产的发展

两晋南北朝时期，四川、湖北、湖南、安徽、江苏、浙江、广东、云南、贵州等省都有茶叶的生产。晋元帝时，有安徽宣城地方官上表贡茶的记载。到了唐代，茶树的种植逐步由内地向长江中下游地区转移，长江中下游地区已经成为当时茶叶生产和技术的中心，这时候，茶叶的产区已经遍及当今中国中南部的14个省区。茶区的分布与近代茶区的分布已经非常接近了。宋代，茶叶传播到全国各地，茶叶的产区范围与现代已经完全相符。宋朝浙江也有郡太守在茶季到茶叶产地监制贡茶的记录。从五代和宋朝的初年起，全国的气候由暖转寒，致使中国南方的茶叶迅速地发展，福建的建安茶成为中国团茶、饼茶制作的技术中心，带动了闽南和岭南茶区的崛起和发展。明清以后，茶区和茶叶的发展主要是体现在六大茶类的兴起，在此之前基本上都是绿茶，或者一些简单的加工，而明清是茶叶加工种类发明最兴旺的一个时期。目前中国有江北、江南、华南和西南四大茶区，有20个省（区、市）自治区，1000多个县（市、区）产茶。中国从茶区面积和茶叶产量来说都是世界第一，茶在中国农业和农村当中有很重要的地位。

（五）茶类的演变和发展

目前得到学术界和业界普遍认可和应用的茶叶分类方法是由安徽农业大学陈椽教授提出来的，依据茶叶加工方法和品质上的差异划分为六大茶类，即绿茶、黄茶、黑茶、红茶、青茶（乌龙茶）、白茶。用这六大基本茶类原料进行再加工形成的茶，比如花茶、紧压茶、萃取茶、茶饮料等属于再加工茶类。

中国制茶历史悠久，在长期的生产实践中逐步总结经验，不断创新，从鲜食到加工，然后加工方法再慢慢演变和发展，西周到东汉到三国再到初唐，从原始的散茶（直接晒干或者烘干）到原始的饼茶（制成饼再烘干）。到了唐宋，又出现了蒸青的团茶和蒸青的散茶，即用蒸汽来完成杀青（多酚氧化酶失活）的过程。到了明朝，烘青发展为炒青，这些都是绿茶的加工方法。黄茶是在杀青后没有及时揉捻和干燥导致茶青闷黄了，加工过程中一次不经意的失误就诞生了另外一个茶类。黑茶起源于16世纪，云南、四川等地的绿茶要通过马帮运输到塞外的新疆、西藏，路途遥远、交通不便，为了节约空间便于运输，就将散茶蒸压成饼茶、砖茶，在漫长的运输途中发生缓慢的后发酵，导致茶的色泽逐步由绿变褐，从而形成了黑茶。红茶是在17世纪的明末清初发明的，茶树鲜叶先经萎凋和揉捻后进行发酵，让多酚类物质被酶充分氧化，产生大量的色素，叶片由绿变为铜红色。最先发明的小种红茶是在福建的武夷山市（原崇安县）的桐木关一带，从小种红茶到工夫红茶，再到后来的红碎茶。青茶又叫乌龙茶。传统的乌龙茶有绿叶红镶边的美称，是因为其独特的摇青工艺，茶青在不断的碰撞摩擦过程中，叶缘细胞破损后氧化变成红色。乌龙茶的生产从福建传向广东和台湾。白茶的加工工序十分简单，即鲜叶摊放，不炒不揉，直接经萎凋干燥成白茶，氨基酸尤其是茶氨酸含量很高。国际贸易上，白茶的需求量也比较大。

二、中国茶的对外传播

世界各国的种茶、饮茶习俗，最早都是直接或者间接从中国传播出去的。这个传播与扩散，经历了一个由原产地到沿长江流域把茶叶传到南方各省，再到韩国、日本、俄罗斯等周边地区，然后逐步走向世界的漫长过程。最初靠马和骆驼通过陆路，后来又开辟了海路。

（一）东传日本

早在公元6、7世纪，日本与中国的佛教交流比较多，最早把茶叶种子带到日本栽种的是日本僧人最澄，他在中国结束学习后回到日本，把茶籽栽种在日本日吉神社的旁边，成为日本最古老的茶园。公元12世纪早期，日本僧人荣西访问中国，他从中国带回更多的茶叶种子，也将中国饮用粉末绿茶的新风俗带回日本，同时也带回了对佛教教义的理解。品茶与佛教理念在日本相互依赖，共同发展，最终成就了一种复杂和独特的仪式，并保存至今，这就是日本茶道。

（二）西传欧洲

1517年，葡萄牙海员从中国带回茶叶。1560年，葡传教士将中国茶叶及饮茶方法等知识传入欧洲。19世纪初期，在英国由一位公爵夫人安娜开发出下午茶并一直流行至今。随着茶叶消费量的不断增加，英国每年需耗费大量资金从中国进口茶叶。

（三）北传俄罗斯

1618年，中国使者带了几箱茶叶到俄国赠送给沙皇。由于路途遥远、行程缓慢，茶叶从中国种植者到达俄罗斯消费者手中需要16~18个月，直到1903年，贯穿西伯利亚的铁路竣工，延续了两百多年的商队贸易才告结束。铁路的开通使得中国的茶、丝绸和瓷器在2个星期内就可以直接运到俄罗斯。

（四）南传印度、斯里兰卡

1780年，印度首次引种中国茶籽；1842年，斯里兰卡开始引种中国茶树。

目前茶叶已行销世界五大洲上百个国家和地区，世界上有50多个国家引种和生产茶叶，160多个国家和地区的人民有饮茶的习俗，饮茶的人口有20多亿。茶叶促进了人类文明的发展，是中国和其他国家，其他民族交流的一个重要桥梁和载体。不仅如此，通过中国茶和茶文化的传播，带动世界茶文化、茶产业的发展。2008年奥运会开幕式上，有一个中华5000年文明的历史长卷，出现了两个汉字，一个是"茶"，一个是"和"，正是借此机会向世界传达中国茶文化的精髓——廉、美、和、敬，由此产生出的文化认同感会让世界更加和谐、有序。

第三节　普洱茶的历史变迁

普洱茶历史悠久，有着浓郁的地方特色和民族文化，下面就详细介绍其历史变迁。

一、普洱茶名称的由来

公元1729年，清雍正皇帝设置普洱府，普洱府治所（官衙）设在现在的宁洱县城里，普洱府下辖现在的普洱市、整个西双版纳州和临沧市部分地区。"普洱"为哈尼语，"普"为寨，"洱"为水湾，意为"水湾寨"，带有亲切的"家园"的含义。普洱一词原是指普洱人，即当今布朗族和佤族的先民濮人，经考证认为：先有普洱人（濮人），后有普洱地名。普洱茶主产于古普洱府所管辖地区，又因自古以来即在普洱集散，因而得名。

二、普洱茶的历史变迁

唐咸通三年（公元862年）樊绰出使云南。在他所著的《蛮书》卷七中有记载："茶出银生城界诸山，散收无采造法。蒙舍蛮以椒、姜、桂和烹而饮之。"银生城即今普洱市景东县城，其管辖范围包括今普洱、西双版纳等地区。但"银生城界诸山"不只局限于银生城，还应包括临沧、大理、德宏、红河、玉溪、保山等地的各大茶山。据考证银生城的茶应该是云南大叶茶种，也就是普洱茶种。所以银生城产的茶叶，应该是普洱茶的始祖。

宋朝的李石在他的《续博物志》一书也记载了："茶出银生诸山，采无时，杂椒姜烹而饮之。"从茶文化的历史认知，茶兴于唐朝而盛于宋朝。中国茶叶的兴盛，除了中华民族以饮茶为风尚外，更重要的因为"茶马市场"以茶叶易换西蕃之马，对西藏的商业交易，开拓了对西域商业往来的前景。

对普洱茶文化来说，元朝是一段非常重要的时期。元朝有一地名叫"步日部"，由于后来写成汉字，就成了"普耳"（当时"耳"无三点水）。普洱一词首见于此，从此得以正名写入历史。没有固定名称的云南茶叶，也被叫做"普茶"逐渐成为西藏、新疆等地区市场买卖的必需商品。普茶一词也从此名震国内外，直到明朝末年，才改叫普洱茶。

明朝万历年间（公元1620年），谢肇淛在他的《滇略》中有记载："士蔗所用，皆普茶也。蒸而成团。"这是"普茶"一词首次见诸文字。明朝，茶马市场在云南兴起，来往穿梭云南与西藏之间的马帮如织。在茶道的沿途上，聚集而形成许多城市。以普洱府为中心点，透过了古茶道和茶马大道极频繁的东西交通往来，进行着庞大的茶马交易。蜂拥的驮马商旅，将云南地区编织为最亮丽光彩的画面。

清朝中叶，古"六大茶山"鼎盛，此时的普洱茶脱胎换骨，变为枝头凤凰，倍受宫廷宠爱更成为贡茶，而且产品远销四川、西藏以及南洋各地，普洱茶从此闻名中外。普洱茶运销之路，就是历史上的茶马古道。

三、普洱茶加工工艺的历史演变

普洱茶以普洱地区命名，千百年来，其独特的韵味享誉古今中外。普洱茶兴于东汉，商于唐朝，始盛于宋，定型于明，繁荣于清。清雍正七年（公元1729年）设置普洱府，下辖今普洱和西双版纳地区，并随着清朝时期普洱茶入贡清政府宫庭受宠而进入发展鼎盛时期。历史上由于交通不便，普洱茶运输只好靠马帮，经茶马古道外运，为便于运输将茶叶制成团、砖、饼等形状之紧茶，且在长途运输、储藏过程中，茶叶产生自然发酵，形成了独特的香气滋味和保健功效。

在20世纪初期，云南普洱茶的生产交易中心是古六大茶山的易武地区和普洱地区，由于战争的原因，各茶庄商号在30—40年代相继歇业，生产跌入谷底，进入了衰退时期。1939年，当时的国民政府派遣范和钧率领一批技术人员到佛海（今勐海）成立官办"中茶公司佛海茶厂"，开云南机械制茶之先河，开始生产机制茶。同样由于战争的原因，茶厂生产跌宕起伏，但云南茶叶生产的中心开始转到代表新技术的佛海地区。

中华人民共和国成立以后，普洱茶生产开始恢复，1952年，佛海茶厂再次复业并在1953年改名为后来的勐海茶厂，成为云南普洱茶中的领军企业。在昆明、勐海、下关三大茶厂没有开始压饼之前，20世纪50年代初出口香港的多是散茶，由于生产已经采取新工艺，到港的晒青毛茶不再有发酵，云南大叶种茶的苦涩、味重、霸气一览无遗，是喝惯了传统发酵普洱茶的港人所接受不了的。另外，中华人民共和国成立之前，也有一些云南茶庄商号的老板到香港定居，带去了传统普洱茶发酵的方法，于是在香港最早开始了用散茶人工发酵加工普洱熟茶，当时有如下几家：联同隆、恒瑞翔、南记、生记、林记、宝泰、同安及长洲福华等。20世纪50—70年代，云南普洱茶多是通过广东省茶叶进出口公司出口到香港的，所以香港人工发酵普洱茶的消息很快就传到了广东，广东也开始研究实验泼水发酵并取得进展。据记载50年代后期，云南也曾实验过热蒸发酵，但没有取得成功。

1973年开始出现了云南普洱茶现代熟茶工艺。根据《云南省茶叶进出口公司志》记载：1973年，云南省茶叶进出口公司派出考察小组对广东的普洱熟茶生产工艺进行了考察，回来后组成技术攻关小组，并最终在昆明茶厂实验成功，从此诞生了现在的渥堆发酵熟茶工艺。

四、"七子饼"的由来

七在中国是一个吉利的数字，七子是多子多福的象征。七子的规制是起自清代，《大清会典事例》载："雍正十三年（公元1735年）提准，云南商贩茶，系每七圆为一筒，重四十九两（合今1.8千克），征税银一分，每百斤给一引，应以茶三十二筒为一引，每引收税银三钱二分。于十三年为始，颁给茶引三千。"这里，清政府规定了云南藏销茶为七子茶，但当时还没有这个提法。

清末，由于茶叶的形制变多，如宝森茶庄出现了小五子圆茶，为了区别，人们将每七个为一筒的圆茶包装形式称为"七子圆茶"，但它并不是商品或商标名称。民国初期，面对茶饼重量的混乱，竞争的压力，一些地区成立茶叶商会，试图统一。如思茅茶叶商会在民国十年左右商定：每圆茶底料不得超过6两，但财大气粗又有政界背景的"雷永丰"号却生产每圆6两五钱每筒8圆的"八子圆"茶，不公平的竞争下，市场份额一时大增。

中华人民共和国成立后，茶叶国营，云南茶叶公司所属各茶厂用中茶公司的商标生产"中茶牌"圆茶。其商标使用年限为1952年3月1日起至1972年2月28日止。因此1970年代初，云南茶叶进出口公司希望找到更有号召力、更利于宣传和推广的名称，他们改"圆"为"饼"，形成了这个吉祥的名称"七子饼茶"。从此，中茶牌淡出，圆茶的称谓也退出舞台，成就了七子饼的紧压茶霸主地位。

茶是中国对世界的贡献。迄今为止，红茶早已红遍世界，茉莉花茶随《好一朵茉莉花》香飘全球，绿茶日益成为世界性健康热点饮品，而普洱茶正成为中国时尚的韵与味，日久而弥香。

第四节　各国茶俗

　　茶叶、咖啡与可可被并称为世界三大无酒精饮料。迄今,世界上有50多个国家种茶,160多个国家的人民有饮茶的习惯。茶,不仅是世界各国人们的生活方式,更是国际交流的重要媒介。现今,国际茶文化交流活动频繁,不同国家茶人常常相聚在一起,共同探讨茶的历史与现状,展望茶文化的未来,在交流中互相学习,相互了解,增进友谊。茶,这样一片树叶,由于各国的国情、民俗不同,历史变迁的差异,造就了丰富多彩的饮用、食用方式,也随之形成了各具特色的茶道文化,也让茶这片树叶在世界各国扮演起了不同的角色。

一、日本茶道

　　日本的历史文化发展与中国有着非常密切的关系,特别在大化革新之后,日本全面吸收唐朝文化体制,与此同时茶和饮茶习惯也就此传入日本。

　　现日本茶道主要包括抹茶道和煎茶道。由于使用的茶不同,器具、手法流程、吃茶方法也都有很大差异。抹茶是把茶的生叶蒸青后干燥,把叶片经石磨碾成极细的茶粉。点茶时将茶粉放入碗中,注入熟汤,用茶筅拂击茶汤至茶水交融,然后分几次喝下。而煎茶则是直接将茶叶放入茶壶中,注入熟汤,再将煎好的茶汤倒入茶碗中饮用。日本煎茶道的形成是受到中国明清泡茶的影响,再加之儒家思想的注入,经过煎茶道始祖"卖茶翁"的统合,最终形成了日本煎茶道的独特思想。

　　"和、敬、清、寂"是日本茶道的中心思想。"和"不仅强调主人对客人要和气,客人与茶事活动也要和谐。"敬"表示相互承认,相互尊重,并做到上下有别,有礼有节。"清"是要求人、茶具、环境都必须清洁、清爽、清楚,不能有丝毫的马虎。"寂"是指整个茶事活动要安静、神情要庄重,主人与客人都是怀着严肃的态度,不苟言笑地完成整个茶事活动。"和、敬、清、寂"始创于村田珠光,正式的确立者则是千利休,400多年来一直是日本茶人的行为准则。

二、韩国茶道

韩国从新罗时期就开始了饮茶之风，至今也有数千年的历史了，而韩国饮茶历史的开启与茶文化的盛行与中国是有着非常密切的联系。

朝鲜《三国本纪》卷十《新罗本纪》中也有专门记载："入唐回使大廉，持茶种子来，王使植地理山。茶自善德王时有之，至于此盛焉。"新罗时代，饮茶盛于寺刹与王室中，这时也出现了茶具与茶器的划分。

高丽时代是饮茶与茶礼的鼎盛时期。这时朝廷专设"茶房"，茶房官员主持宗庙祭祀、接待外国使臣、国王外出时用茶准备；司宪府中每天有规定的"茶时"，公务人员喝茶后头脑清晰，可确保公务行之有效，提倡"饮茶乃清白官吏养成美德之途径"；僧侣们不但在各种礼佛仪式上用茶，在日常修行生活中也经常饮茶；饮茶风在市民阶层也普及了，城中设有"茶房、茶店"，专门售卖茶叶、茶水。

到朝鲜时代，出现了饮茶生活走向衰弱，而茶文化精神走向顶峰的交叉时期。此间涌现出了大批优秀的茶专著与作品，如草衣禅师撰写了与陆羽《茶经》比翼的《东茶颂》，后人把草衣禅师尊为韩国的"茶圣"。

韩国茶道追求的是"中正"精神。在草衣禅师的《东茶颂》中，两次提到了"中正"一词，"體神雖全猶恐過中正，中正不過健靈併"。1979年，在韩国茶人会上，正式把"中正"确立为韩国茶道文化的中心思想，并以此为茶人的行为准则。"中正"包含了四层含义，既不多余，又不缺少；万人平等；先人后己；追思根源，回归自然。

三、英式下午茶

1662年，葡萄牙公主嫁给英国国王查理二世，中国茶作为她的嫁妆带到了英国。由于凯瑟琳的宣传提倡，饮茶之风盛行于英国宫廷，又扩展到王公贵族及富豪、世家，后来饮茶进一步普及到民间大众，风靡英国。

"当时钟敲响四下时，世上的一切瞬间为茶而停。"这是英国一首著名的民谣，饮茶可以说就是英国人的一种生活方式。据统计，英国每年消耗近20万吨的茶叶，占世界茶叶贸易总量的20%，有80%的英国人每天有饮茶的习惯，年人均饮茶约3千克，茶叶消费量几乎占各种饮料总消费量的一半，由此可见，英国确是名副其实的"饮茶王国"。

英国人最初是饮用绿茶和武夷茶，后来渐渐改变成喝红茶，并且发展出自己独特的红茶文化。英国人每天的饮茶时间非常多，寝觉茶、早餐茶、早休茶、午餐茶、午休茶、下午茶、餐后茶等等，即使在忙碌的上班时间也要抽空来喝杯茶。茶已成为英国人生活中不可或缺的重要元素。

下午茶是在18世纪就存在的饮茶习惯，最早是由贝德芙公爵夫人安娜女士所开创的。一般来说下午茶是一个喝茶的时段，由于午餐吃得简单，晚餐又吃得较晚，相隔的时间较长，因此选择在四点这个时段喝茶，吃些小点心。主人会把家中珍藏的茶具拿出来，铺上美丽的桌巾，放上点心，香醇的红茶招待左邻右舍或亲朋好友，下午茶可以说是英国人社交活动的重要途径之一，是人际交往的重要桥梁。

四、俄罗斯人的茶饮生活

俄罗斯人第一次接触茶是在1638年。当时作为友好使者的俄国贵族瓦西里·斯塔尔可夫遵沙皇之命赠送给蒙古可汗一些紫貂皮，蒙古可汗回赠的礼品便是4普特（约64千克）的茶。品尝之后，沙皇即喜欢上了这种饮品，从此茶便登上皇宫宝殿，随后进入贵族家庭。

从17世纪70年代开始，莫斯科的商人们就做起了从中国进口茶叶的生意。清朝康熙皇帝在位时，中俄两国签订了关于俄国从中国长期进口茶叶的协定。但是，从中国进口茶叶，路途遥远，运输困难，数量也有限。

因此，茶在17、18世纪的俄罗斯成了典型的"城市奢侈饮品"。直到18世纪末，茶叶市场才由莫斯科扩大到少数外省地区，到19世纪初饮茶之风在俄国各阶层才开始盛行。

俄罗斯人酷爱喝红茶，喝红茶时习惯于加糖、柠檬片、果酱、奶油，有时也加牛奶。这是为了让红茶中充满果香、甜香，去除红茶中的涩味。

俄罗斯人饮茶不能不提到有名的俄罗斯茶炊，有"无茶炊便不能算饮茶"的说法。茶炊实际上是喝茶用的热水壶，装有把手、龙头和支脚。茶炊的四周装水和茶叶，中间有一个空筒，可放燃烧的木炭，金属外壁安有小水龙头，烧开后可直接放茶水。在现代俄罗斯家庭，很多人更多地使用电茶炊。电茶炊的中心部分已没有了盛木炭的直筒，也没有其他隔片，茶炊的主要用途变成单一的烧水。人们用瓷茶壶泡茶叶，3~5分钟之后，往杯中倒入适量泡好的浓茶水，再从茶炊里接煮开的水入杯。

虽然旧时的传统茶炊已渐渐变为装饰品、工艺品，但每逢佳节，俄罗斯人还是会把茶炊摆上餐桌，家人朋友围坐在茶炊旁饮茶，其乐融融的氛围随着一杯茶传递着。

/ 第二章 /

普洱茶 基础知识

第一节 茶叶的分类

茶叶是以茶树［*Camellia sinensis* (L.) O.kunts］的芽、叶、嫩茎为原料,以特定工艺加工、不含任何添加剂、供人们饮用或食用的产品。中国是茶的发源地,也是茶类最多最齐全的产茶国家,茶叶的品类繁多,命名复杂。我国茶叶的分类方法较多,有以销路分类,有以制法、品质或季节等分类。安徽农大茶学专家陈椽教授提出六大茶类分类方法,并建立了六大茶类分类系统。中茶所程启坤研究员提出茶叶综合分类方法,将茶分为基本茶类(即六大茶类)和再加工茶类(花茶、紧压茶、含茶饮料等)两大类的基本分类法。该方法至今为国内外所接受。本书中又增加一种茶叶分类方法:"三位一体"分类法。

一、茶叶综合分类法

此种方法以生产工艺、产品特性、茶树品种、鲜叶原料和生产地域进行分类。大体上将茶叶分为基本茶类和再加工茶类两大部分。根据制造方法不同和品质上的差异,通常将茶叶分为绿茶、红茶、青茶(即乌龙茶)、白茶、黄茶和黑茶六大类。

1.绿茶

绿茶属不发酵茶。是鲜叶采摘后经杀青、揉捻、干燥而成的茶。以杀青和干燥方式进行分类,又可分为:炒青绿茶,如西湖龙井、碧螺春;蒸青绿茶,如湖北的恩施玉露、日本的煎茶;烘青绿茶,如黄山毛峰、太平猴魁;晒青绿茶,如滇青、川青。

绿茶

2. 红茶

红茶属全发酵茶。制作过程是萎凋、揉捻、发酵和干燥。分为小种红茶、工夫红茶和红碎茶三大类。

红茶

3. 青茶（乌龙茶）

青茶属半发酵茶。青茶主产于福建、广东和台湾，可分为：闽南乌龙，如铁观音、黄旦、本山、梅占、毛蟹；闽北乌龙，如武夷岩茶、水仙、肉桂等；广东乌龙，如凤凰单丛、凤凰水仙；台湾乌龙，如冻顶乌龙、文山包种、白毫乌龙等。

乌龙茶

4. 白茶

白茶属微发酵茶。鲜叶采摘下来后经过萎凋后，直接晒干、烘干或阴干而成。主产地在福建和云南一带，如福建的白毫银针、白牡丹、寿眉，云南的月光白等。

白茶

5. 黄茶

黄茶属部分发酵茶。在制作过程中需经闷黄，使茶叶与茶汤的颜色呈黄色。黄茶种类有：黄芽茶，如君山银针、蒙顶黄芽等，此外还有黄小茶和黄大茶。

黄茶

6. 黑茶

黑茶属后发酵茶。基本加工工序包括杀青、揉捻、渥堆、干燥等。如云南普洱茶、湖南安化黑茶、湖北老青茶、四川康砖茶、广西六堡茶等。

再加工茶主要有花茶、紧压茶和萃取茶。花茶如茉莉花茶、珠兰花茶、玫瑰花茶等；紧压茶如黑砖、饼茶等；萃取茶叶叫速溶茶。

黑茶

这种将茶叶分为基本茶类和再加工茶类的分类方法是中国一直采用的茶叶分类方法，具体分类信息见图2-1茶叶综合分类法。

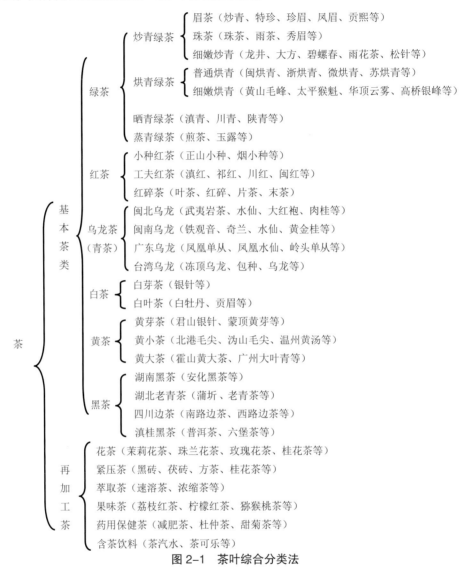

图 2-1　茶叶综合分类法

二、"三位一体"分类法

20世纪90年代茶叶深加工技术水平不断提高，茶新产品不断涌现。针对这种状况西南大学茶学专家刘勤晋教授提出了茶"三位一体"分类方法，即茶和茶制品的分类见图2-2：将茶分为茶叶饮料、茶叶食品、茶叶保健品、茶叶日用化工品及添加剂，其中茶叶饮料又分泡饮式、煮饮式、直饮式三类。

```
                    ┌ 绿茶——按制法分炒青如珍眉、珠茶、龙井、贡熙；烘青如黄山毛峰、
                    │        四川蒙顶甘露；晒青如川青、滇青；蒸青如玉露
                    │ 黄茶——按制法分湿坯闷黄：远安鹿苑茶、蒙顶黄芽、台湾黄茶；干
                    │        坯闷黄：君山银针、安溪黄大茶
                    │ 青茶——按制法分筛青做青：闽北水仙、大红袍、武夷岩茶；摇青做
                    │        青：台湾包种、冻顶乌龙、安溪铁观音
              ┌ 泡饮式┤ 花茶——按窨花种类分茉莉花茶、朱兰花茶、玳玳花茶、玫瑰红茶、
              │      │        中国香兰、荔枝红茶
              │      │ 红茶——按制法分工夫红茶如祁门红茶、川红工夫；小种红茶如正山
              │      │        小种、烟小种；红碎茶如C.T.C红茶、洛托凡红茶、转子机
              │      │        红茶
              │      │ 白茶——按嫩度分芽茶如白毫银针、叶茶如白牡丹、贡眉等
              │      └ 黑茶——按产地分云南普洱茶、安化黑茶、广西六堡茶、日本阿波晚
              │               茶、苏联老青茶
              │      ┌ 砖茶（紧压茶）——按形状可分为篓装紧压茶如四川康砖、金尖、方
      ┌ 茶叶饮料┤      │        包；砖形茶如湖南黑砖、花砖、茯砖、湖北老青砖、紧茶、
      │        │      │        云南方茶
      │        │      │ 腌茶——云南竹筒茶、日本富山黑茶、其石茶
      │        │ 煮饮式┤ 擂茶——湖南擂茶、广西擂茶
      │        │      │ 油茶——湖南湘西油茶、四川土家油茶、西藏酥油茶、蒙古奶茶
      │        │      └ 速溶茶——速溶红茶、速溶绿茶、速溶乌龙茶、速溶花茶、速溶普洱茶
      │        │      ┌ 冰茶——柠檬冰茶、苹果冰茶、香草冰茶、麦香冰茶
      │        │      │ 汽水茶——柠檬汽水茶、荔枝汽水茶、香草汽水茶、果味汽水茶
      │        └ 直饮式┤ 泡沫茶——泡沫红茶、泡沫乌龙茶、泡沫包种、泡沫铁观音
      │               │ 茶头罐头——荔枝红茶、麦香红茶、乌龙茶、玉露绿茶、茉莉花茶
      │               └ 茶酒——四川茶酒、日本茶酒
茶┤            ┌ 茶糖果——红茶奶糖、红茶朱古力、绿茶饴
      │ 茶叶食品┤ 茶点心——红茶饼干、红茶蛋糕、浆茶三明治、绿茶馒头
      │        │ 菜肴——龙井虾仁、樟茶鸭子、清蒸茶鲫鱼、绿茶番茄汤、凉拌嫩茶尖
      │        │ 茶饭——茶粥、鸡茶饭、盐茶鸡蛋
      │        └ 茶冷饮品——红茶冰淇淋、红茶娃娃糕、绿茶冻
      │        ┌ 茶多酚——维多酚、儿茶酚口服液
      │ 茶叶保健品┤      ┌ 专用药茶：宁红保健茶、上海保健茶、清咽保健茶、防龋茶、降
      │        │ 保健茶┤          糖茶、降脂茶
      │        │      └ 补药茶：人参茶、富硒茶、杜仲茶、八珍茶、参芪茶
      │        └ 茶多糖抗辐射制剂（针剂）
      └ 茶叶日用化工品及添加剂┤ 茶叶抗氧化剂、茶叶色素、茶叶保鲜剂
                         └ 茶皂素制品：茶叶洗发香波、茶叶防臭剂、茶叶起泡剂（表面
                                     活性剂）
```

图2-2 茶和茶制品的分类

第二节　普洱茶基础知识

一、什么是"普洱茶"？

2008年，中华人民共和国国家质量监督检验检疫总局和中国国家标准化管理委员会联合发布了中华人民共和国国家标准《地理标志产品 普洱茶》（GB/T 22111—2008），并于12月1日起正式实施。目前，普洱茶的定义一般以GB/T 22111—2008为准。该标准在对普洱茶的产地范围、茶树品种、加工工艺、品质特征和分类等均作了明确规范要求：以地理标志保护范围内的云南大叶种晒青茶为原料，并在地理标志保护范围内采用特定的加工工艺制成，具有独特品质特征的茶叶。按其加工工艺及品质特征，普洱茶分为普洱茶（生茶）和普洱茶（熟茶）两种类型，按外观形态分普洱茶（熟茶）散茶、普洱茶（生茶、熟茶）紧压茶。

该标准规定普洱茶地理标志产品保护范围是：云南省普洱市、西双版纳傣族自治州、临沧市、昆明市、大理白族自治州、保山市、德宏傣族景颇族自治州、楚雄彝族自治州、红河哈尼族彝族自治州、玉溪市和文山壮族苗族自治州等11个州（市），75个县（市、区）所属的639个乡镇。只有在地理标志保护范围内的特定原料并且在地理标志保护范围内生产加工的才能称为普洱茶。此外，该标准还首次定义了普洱茶的后发酵工艺：云南大叶种晒青茶或普洱茶（生茶）在特定的环境条件下，经微生物、酶、湿热、氧化等综合作用，其内含物质发生一系列转化，而形成普洱茶（熟茶）独有的品质特征的过程。

本书中，将"普洱茶（生茶）"简称为"普洱生茶"或"生茶""普洱茶（熟茶）"简称为"普洱熟茶"或"熟茶"。

二、普洱茶的特点

普洱茶作为一种特有的茶类，它有与其他茶类截然不同的特点，也就是普洱茶的"六奇"：

1.产地奇：经历了由地名命名而发展为专门茶类的一种茶叶，地域特征明显。普洱茶的原产地在澜沧江下游一带的各大茶山，尤其以六大茶山为代表，超出了这特定区域范围的茶叶严格上讲不能称之为普洱茶；

2. 品种奇：原料为云南大叶种茶树鲜叶，大叶种茶不同于中小叶种茶，其外形上是乔木大树，叶片较中小叶种大，且茶芽肥厚柔嫩，内含物质丰富，活性成分高，保健功效强；

3. 制作工艺奇：普洱茶是用云南大叶种晒青茶为原料，经过后发酵制作而成；其制作工序是：云南大叶种鲜叶→杀青→揉捻→晒干→晒青毛茶（蒸压成型成为普洱生茶）→渥堆→翻堆→干燥→分筛→拣剔→拼配（普洱熟茶散茶）→蒸压做形→干燥→普洱熟茶（紧压茶）。

4. 形状奇：普洱茶除散茶外，紧压成型的普洱茶有各种形状，小的如3克迷你沱茶、100克沱茶、250克沱茶、砖型、圆饼，大的如金瓜、葫芦、屏风、大匾等。

5. 品质奇：普洱茶属于后发酵茶，有越陈越香的特点，适宜条件下存放的时间越久，品质越好，价格越贵，也越受老茶人的喜爱和追捧。如贮存保管得当，可贮存几十年，因此被称为能喝的"古董茶"，将茶视为古董，唯有普洱茶。

6. 饮用奇：普洱茶是最讲究冲泡技巧和品饮艺术的茶类，在云南，普洱茶的冲泡非常讲究，对水、器等都有研究，这个在后面的品鉴里会详细介绍，除了清饮外，人们喜欢把普洱茶和自己民族的传统和风俗结合起来形成多样的品饮方法，制作成各种调饮茶。

正是因为上述特点，使普洱茶成为可收藏、鉴赏的"古董"，世界上除了法国波尔多的红酒外，就是中国云南的普洱茶被称为可以喝的"古董"，这是大多茶类都无法具备的特性。

三、普洱茶的分类

普洱茶就像一本永远翻不完的书，每翻一页，让人总有摸不着头脑之"名词"。究其原因，普洱茶市场关于普洱茶的言论太多、太杂。"乔木茶""大树茶""古树茶""野生茶""干仓茶""沱茶"等等，何解？细致梳理梳理，从分类上理解，掌握其规律，不仅可以理清头绪，还得以看穿虚假，回归本质。

（一）按外形分类

普洱茶散茶：制茶过程中未经过紧压成型，茶叶为散条型的普洱茶散茶，普洱散茶按品质特征分为特级、一级至十级共十一个等级。

（二）普洱茶紧压茶

1. 饼茶：扁平圆盘状，其中七子饼每块净重357克，每7个为一筒，每筒重2500克，故名七子饼。

2. 沱茶：形状跟碗一般大小，有100克、250克等，还有迷你小沱茶每个净重2~5克。

3. 砖茶：长方形或正方形，250~1000克居多，制成这种形状主要是为了便于运送。

4. 金瓜贡茶：也称团茶、人头贡茶，是普洱茶独有的一种特殊紧压茶形式，因其形似南瓜，茶芽长年陈放后色泽金黄，得名金瓜。早年的金瓜茶是专为上贡朝廷而制，故名"金瓜贡茶"，从100克到数百斤均有。

市场上还可根据个人喜好定制成各种形状的普洱茶。

（三）按加工方法分类

普洱茶原料的制作工序是：云南大叶种茶树鲜叶经过杀青→揉捻→晒干→晒青毛茶。根据对晒青毛青的加工方法不同，普洱茶分为生茶和熟茶两大系列。

1. 普洱生茶：是以地理标志保护范围内生长的云南大叶种茶树鲜叶为原料，经杀青、揉捻、日光干燥、蒸压成型等工艺制作的紧压茶。其品质特征为：外形色泽墨绿、香气清纯持久、滋味浓厚回甘、汤色绿黄清亮，叶底肥厚黄绿。

2. 普洱熟茶：是以地理标志保护范围内生长的云南大叶种茶树鲜叶制成的晒青毛茶为原料，采用渥堆发酵（微生物固态发酵）等特定工艺加工形成的的散茶和紧压茶。其品质特征为：外形色泽红褐，内质汤色红浓明亮，香气独特陈香，滋味醇厚回甘，叶底红褐。

（四）按茶树树龄分类

加工云南普洱茶的茶树类型主要为乔木型茶，根据制茶茶树树龄的不同又可分为小树茶、中树茶、大树茶和古树茶。

1. 小树茶：小树茶指制作原料为种植年限30年以下。

小树茶

2. 中树茶：中树茶指制作原料为种植年限30年以上、60年以下。

3. 大树茶：大树茶指制作原料为种植年限60年以上、100年以下。

4. 古树茶：泛指制作原料为种植年限100年以上的古茶树。古树茶获取土壤深层的矿物质成分，能以内质丰富的最佳状态将各山头的独特性体现出来。此类原料制成的普洱茶为"饮茶发烧友"追捧，古树茶原料有限，市面上价格较高，但其独特内质更能体现普洱茶的"茶文化"。

另外根据是否为人工栽培将茶树分为栽培型茶树和野生型茶树，现在市场上主要是栽培型茶树。

中树茶

大树茶

古树茶

（五）依存放方式分类

1. 干仓普洱茶：干仓普洱茶是指存放通风、干燥及空气湿度小（一般指空气湿度小于70%）的清洁仓库环境里，使茶叶发生自然后发酵。

2. 湿仓普洱茶：通常存放于高温高湿的地方，可以加快其发酵速度。对待"湿仓"茶，首先应消除"湿仓茶就是老茶"的误解，湿仓茶虽然陈化速度更快，但是茶叶在高温高湿环境下难免会发霉。而肉眼可见的发霉茶，嗅觉可闻到的刺鼻茶，品茶中喉、舌、口感觉到叮、刺、挂的锁喉茶，无论是从健康、保健角度，还是同新世纪提倡的绿色无污染、生态、有机食品要求都是背道而驰的，所以我们不主张销售及饮用湿仓普洱茶。

四、普洱纯料茶与拼配茶的区别

普洱茶一直有"纯料"与"拼配"的争议。一种观点认为，普洱茶"纯料"或"一口料"的原料选用，是传统普洱茶制作的基本规则；另一种观点则认为，普洱茶的"拼配"是品质再优化和再提高的一种工艺手段，经典的普洱茶产品一定有其独特的"保密配方"，而这个配方的核心内容就是"拼配"。以下我们分几个方面来明确如何正确的理解普洱茶的纯料与拼配。

（一）纯料茶

目前市场对纯料茶的一些认定分类，主要有五种形式：

第一种古树纯料，不同山头的古树茶。这种"纯料"茶的特点就是只要是古树茶，不管你是哪个山头的，也不管你是春茶还是秋茶，就是纯料。

第二种季节纯料，同一个季节的茶。这种纯料不分茶树的树龄大小，也不分批次，只分季节，也就是说只分春茶、夏茶和秋茶。

第三种同一茶园同一批次纯料，同一片茶园的古树茶，不分批次，不分茶树树龄大小。这种纯料茶的特点就是同一片茶园，而不是同一个山头或者是寨子，但是也是不分采摘批次的。

第四种同一茶园不同批次纯料，同一片茶园里的茶，分树龄、分批次。这种分法就比前面的严谨多了，这样的纯料必须是差不多一起采摘的而且树龄也差不多，因此这样的纯料量相对较少。

第五种单株纯料,"单株"又称"一棵树"。意思就是"单株茶",也就是单一古树上的茶。一般都是选择古茶园里茶树树龄最大的那一棵茶树,或者是几棵茶树,只采摘头春茶,所以量是很少的,一般都是私人珍藏品,很少有在市场上出售的,这种茶是发烧友的最爱。

(二)拼配茶

普洱茶的"拼配"涵盖六个方面:等级的拼配,不同茶山的拼配,不同品种的拼配,不同季节的拼配,不同年份的拼配,不同发酵度的拼配。普洱茶等级拼配是普洱茶拼配中最常见的一种方法。无论是新茶,还是年份较长的老茶,其底、面、内的茶叶等级都有差别。即使这种差别很小,都有独特的拼配技术在里面,一饼普洱茶的"层次感"离不开等级拼配的技巧。因此普洱茶的拼配自古至今是广泛存在的。

普洱茶的拼配注重的是茶叶内含物质的"优势互补"。不同茶山、不同区域所生产的晒青毛茶其口感差异很大。这种差异不仅体现茶叶芳香类物质含量的不同,还体现了内含物一些细微差别。如何判断这种差异,以及将这种差异进行有效合理的"重组"与"融合",创造一种更优质的普洱茶产品,是从古至今普洱茶人始终追求的梦想。普洱茶的拼配可形成普洱茶后续发酵的梯级转化。以饼茶后发酵为例，它要求压制的饼茶松紧适度，即不能太密实（紧压过度），又不能太松弛（间隙太大）。解决这一问题最好的方法是不同等级原料的合理拼配，以七级茶做"骨架"，以三级或五级茶"添实补缺"。这种网状骨架的搭建，可使普洱茶出现层次感，并使后续的发酵出现梯级转化。很多人对普洱茶的品级存在一个误区，认为所选用的毛料级别，越高越好，甚至有人追捧纯芽头（特级）制成的饼茶，且误认为普洱茶的原料级别越高，营养价值越高，反之，级别越低，营养物质越少。以普洱茶原料中还原总糖含量为例，权威部门检测的结果是七级茶含量最高。这正是七级茶被大量用于饼茶的主要原因，这种粗老茶叶不仅是饼茶形成"网状骨架"的主力，同时也因内含物质的特性，使它成为普洱茶后续发酵与转化的"骨干力量"。

普洱茶的拼配是一种极具个性化色彩的艺术。凡是经典的普洱茶产品，无论是流传几十年的老茶，还是近几年的"新品"，都有各自独特的"茶性"，只要我们深入体会，都会找到它们的差别，哪怕是一些细微的差别。这种差异化具有浓重的个性色彩，会使我们的味觉产生深刻的记忆而久久不忘。这种感觉，或者说是品质，不是简单的"纯料"和"一口料"所能赋予的。更多的是普洱茶制作者长年的经验总结和感悟中的智慧结晶，是高超的拼配技艺结出的"硕果"。

勐海悦成纳达勐茶园

第三章 普洱茶保健功效

第一节　普洱茶降脂减肥功效

现代医学证实,普洱茶有显著降脂减肥的功效,且有美容和延年益寿的功能。随着饮用者人数日益增加,普洱茶备受关注,有"苗条茶"和"延年茶"的美称。目前饮用普洱茶已改变了传统意义上的饮茶解渴,而是根据普洱茶特性将之视为良药,它与健康长寿,降脂减肥等功效联系在一起,这给普洱茶增加了新的内涵。

一、肥胖症概述

（一）肥胖症概念

肥胖症又称作肥胖病,由于遗传性因素与环境作用,引起的营养代谢障碍性疾病,最明显的特征是机体摄入能量大于消耗能量,引起体内脂肪聚集过多的症状。当前肥胖问题已成为全球公共卫生问题关注的对象,国际肥胖特别工作组指出,肥胖将成为威胁人类健康的最大杀手。

（二）肥胖症的特征

个体肥胖主要表现出脂肪细胞数量增大和体积增加。一般而言,当一个人的体重超过标准体重的20%以上,或身体质量指数(body mass index,简称体重指数,BMI)大于$24kg/m^2$时,就称为肥胖症。

（三）肥胖症产生的原因

引起肥胖症发生的原因虽说有许多种,但最基本的一条就是体内能量代谢平衡发生失调。许多因素都可以导致患者体内能量代谢发生障碍(失调),如营养过剩、体力活动减少、内分泌代谢失调、下丘脑损伤、遗传因素或情绪紊乱等都可能导致肥胖症的发生。

（四）肥胖症的危害

肥胖影响了体态,并且对身体健康有害,肥胖会引起许多慢性疾病,如心血管疾病相关的多种代谢功能异常,增加Ⅱ型糖尿病、冠心病、高血压、中风、充血性心力

衰竭、脂质异常血症、睡眠时呼吸暂停综合症及某些癌症（如卵巢癌、胸腺癌和结肠癌）等的发病率和死亡率。肥胖是人们健康长寿的天敌，科学家研究发现肥胖者并发脑栓塞与心衰的发病率比正常体重者高1倍，患冠心病、高血压、糖尿病、胆石症者较正常人高3~5倍，由于这些疾病的侵袭，人们的寿命将明显缩短。身体肥胖的人往往怕热、多汗、皮肤皱褶处易发生皮炎、擦伤，并容易合并化脓性或真菌感染；而且由于体重的增加导致身体各器官负担加重，容易遭受各种外伤、骨折及扭伤等。此外，睡眠呼吸暂停综合症、恶性肿瘤的产生等都与肥胖也有着直接的关系。

二、普洱茶对肥胖症的作用

关于普洱茶的减肥作用，最早研究的是日本学者Mitsuaki Sano，他在1985年的试验证明给高脂大鼠饲喂普洱茶，可以降低高脂大鼠血管内的胆固醇和甘油三酯含量，显著降低高脂大鼠腹部脂肪组织重量。随后，Yang在1997年也报道，高胆固醇造模后的大鼠在饲喂普洱茶后，食物和饮水消耗减少，体重下降，血液和肝脏中的胆固醇和甘油三酯含量下降，高密度脂蛋白胆固醇含量增加。2005年，Kuo等人的试验结果表明，正常大鼠喂饲普洱茶30周后，体重、胆固醇和甘油三酯含量均显著降低，且降低幅度大于其他茶类如绿茶，乌龙茶和红茶，同时低密度脂蛋白胆固醇降低，而高密度脂蛋白则显著升高，抗氧化酶SOD活性较正常对照组要高。同时，国内研究人员也报道了喂食高脂饲料的小鼠在同时喂食晒青毛茶或普洱茶时，均能有效地抑制高脂饮食小鼠血脂的升高，并能使血清TG、TC、LDL-C水平全面降至正常值范围，同时使高密度HDL-C水平显著升高，普洱茶的效果优于晒青毛茶。

熊昌云利用动物基础饲料（M02-F）和高脂饲料（M04-F），按照M04-F与2.5%、5%、7.5%的普洱茶熟茶粉分别配制成低、中、高3个剂量的含茶高脂饲料，搅拌机拌匀后喂养供试大鼠。通过空白对照组、肥胖模型对照组、饮食控制组、普洱茶低剂量组、普洱茶中剂量组和普洱茶高剂量组的对比实验研究，发现实验结束时，相对于肥胖模型组来说，饮食控制组大鼠体重有明显下降的趋势，达到显著差异（$P<0.05$）；在普洱茶的3个剂量处理组中，中、高剂量组大鼠体重也有明显的下降，其作用接近于饮食控制组；低剂量普洱茶组则没有表现出对肥胖大鼠体重增长的抑制效果。另一方面，有趣的是，各处理组大鼠摄食量并没体现出显著性差异，说明经过不同剂量普洱茶处理后，肥胖大鼠体重增量的减少不是通过对食物的摄入量减少而引起的。血清总胆固醇（TC）和甘油三酯（TG）含量是评价大鼠肥

胖的重要指标。实验结束时，肥胖模型组大鼠的血清TC、TG水平都显著高于空白对照组（$P<0.01$）。普洱茶处理前，肥胖，模型组和各实验处理组大鼠的TC水平是一致的，处理后，普洱茶3个剂量处理组大鼠TC水平相对于肥胖模型组都有明显的下降，达到了显著差异水平（$P<0.05$）；饮食控制组大鼠的TC水平下降则表现出了极显著差异（$P<0.01$）。同时，相对于肥胖模型组而言，普洱茶中、高剂量处理组大鼠的血清TG水平分别下降了20.10%和25.62%。普洱茶高剂量处理组大鼠的血清TG水平已接近饮食控制组的TG水平（27.21%）。这些实验结果显示普洱茶和饮食控制都能有效降低营养肥胖性大鼠血清TC和TG含量，改善大鼠的血清指标质量，对大鼠肥胖症的预防或治疗肥胖症有着潜在的价值和意义，而且高剂量普洱茶处理的效果接近于饮食控制的效果，这将为那些为了减肥而特意控制饮食的人们提供了一条利用普洱茶来代替控制饮食减肥的途径。

高密度脂蛋白胆固醇（HDL-C）是血清总胆固醇的一个主要的组成部分，被视为动物体内的"好胆固醇"。熊昌云实验结果证明，与肥胖模型组比较，普洱茶和饮食控制处理均能有效提高营养肥胖性大鼠体内血清HDL-C的含量。经过普洱茶处理6周后，低、中、高剂量处理组肥胖大鼠体内血清HDL-C含量分别增加了27.59%（$P<0.05$）、43.68%（$P<0.01$）和62.07%（$P<0.01$），饮食控制组仅增加了21.84%（$P<0.05$）。这表明普洱茶对肥胖性大鼠HDL-C指标的提高效果要优于单纯的饮食控制，且表现出剂量效应，最高剂量普洱茶处理组对提高肥胖大鼠HDL-C水平表现出了最好的效果。而更令人欣喜的是，最高剂量普洱茶组大鼠的HDL-C的水平远超过空白对照组，达到了极显著差异水平（$P<0.01$）。

动脉粥样硬化指数（AI）是由国际医学界制定的一个衡量动脉硬化程度的指标。肥胖性大鼠在普洱茶不同剂量和饮食控制的处理下,其AI值下降非常明显,与肥胖模型组相比，普洱茶低、中、高剂量处理组分别下降了57.38%、69.20%、79.75%，饮食控制组的下降率则为62.87%，均体现出极显著差异（$P<0.01$）。而相对于空白对照组而言，饮食控制组的AI值已与其持平，说明通过饮食控制可以使肥胖性大鼠动脉粥样硬化程度恢复到以前的水平，而普洱茶中、高剂量处理组的AI值则低于空白对照组，尤其是高剂量普洱茶处理组与空白对照组相比达到了极显著差异。这样的结果表明普洱茶在抗动脉粥样硬化方面有着显著效果，不仅能抑制由于摄入过量高脂饲料引起的肥胖大鼠AI值的上升，而且还能改善正常大鼠的血清指标，降低AI值,减少动脉粥样硬化风险，其作用效果是单纯的饮食控制所不能达到的。

第二节　普洱茶抗疲劳与抗衰老功效

一、疲劳和衰老概述

随着社会的进步，医学模式的转变，健康的概念也发生着转变。抗疲劳与抗衰老成了人们关注的问题。

临床上，疲劳是亚健康的一种常见表现。然而，疲劳是一个非常普遍的症状或现象，不仅可存在于健康人群、亚健康人群中，且许多疾病人群也常常存在疲劳症状。疲劳产生大致有三方面的原因：一因持续做功，超过机体所能承受的能力所致；二因某些负面情绪引起；三因疾病造成。疲劳是多种原因所致的局部组织、器官功能减退或全身不适的主观感觉，有一过性疲劳和累积性疲劳。疲劳的表现可体现在躯体方面，如体力减退感、无力感；也可体现在精神方面，如表现为对活动（体力或脑力的）的厌恶感。在行为学上表现为工作效率的下降。

衰老又称老化，是机体各组织、器官功能随着年龄增长而发生的退行性变化，是机体诸多生理、病理过程和生化反应的综合体现，是体内外各种综合因素（包括遗传、营养、精神因素、情绪变化、环境污染等）共同作用的结果。衰老是人类生命发展中的必然趋势，是不以人的意志为转移的客观规律，任何人都不能阻止衰老的进程，但可以通过科学的方法延缓其进程。

人体衰老的外部特征表现主要有：①皮肤松弛发皱，特别是额及眼角。②毛发逐渐变白而稀少。③老年斑出现。④齿骨萎缩和脱落。⑤骨质变松变脆。⑥性腺及肌肉萎缩，出现"更年期"的各种症状，例如女人的经期紊乱、发胖；男人发生忧郁、性亢进、失眠等。⑦血管硬化，特别是心血管及脑血管的硬化和肺及支气管的弹力组织萎缩等。

功能特征表现主要有：①视力、听力减退。②记忆力、思维能力逐渐降低。③反应迟钝，行动缓慢，适应力低。④心肺功能下降，代谢功能失调。⑤免疫力下降，因此易受病菌侵害，有的还产生自身免疫病。⑥出现老年性疾病，如高血压、心血管病、肺气肿、支气管炎、糖尿病、癌肿、前列腺肥大和老年精神病等。

二、普洱茶的抗疲劳、抗衰老作用

慢性疲劳已成为困扰人们正常工作和生活的一种疾病现象。长期以来，众多学者期望能寻找到一种安全、有效、无毒副作用的良方来延缓疲劳的发生和加速疲劳的消除，而茶素有"提神解乏，明目利尿，消暑清热"的功能，具有广阔的开发前景。张冬英等选用具有代表性的普洱茶熟茶样品，通过动物小鼠模型探讨普洱茶的抗疲劳效果。选用勐海县云茶科技有限责任公司、云南龙润茶业集团和云南省思茅茶树良种场生产的普洱茶熟茶，将3个供试茶样等量混合作为受试物。茶样经沸水浸提、抽滤机过滤、合并浸提茶汤、减压浓缩、茶汤浓缩液装瓶灭菌，制备得到1克/毫升的茶汤浓缩液。

将小鼠按体重随机分为4个剂量组：空白对照组和普洱茶熟茶低、中、高剂量组，每组20只小鼠，雌雄各半。实验采用昆明种小鼠按人体推荐量的5倍作为最低剂量，中、高剂量分别为人体推荐量的10倍和20倍，即低、中、高剂量组分别给予普洱熟茶0.5、1.0、2.0克/千克；空白对照组给予生理盐水（0.9%）。每周称重1次，每天早上（9:00—11:00）根据体重经口灌胃给药1次，连续给药30天。进行小鼠负重游泳试验和生理生化指标血乳酸（BLA）、血尿素氮（BUN）、血乳酸脱氢酶（LDH）及肝糖元（LG）、肌糖元（MG）的测定。实验结果表明，普洱茶熟茶低、中、高剂量组的小鼠在实验结束时的平均体重与实验开始时的平均体重相比，分别增加了25.22%、28.03%和18.17%，阴性对照组则增加了31.37%。从外观上看，高剂量组小鼠体型较为瘦长。表明低、中、高剂量的普洱茶熟茶对小鼠体重的增长均具有显著的抑制作用，且以高剂量效果最佳。小鼠负重游泳时间是抗疲劳作用的直接反应，与抗疲劳效果成正相关。与空白对照组相比，普洱茶熟茶低、中、高剂量组的小鼠负重游泳时间均有所延长，增加率分别为25.74%、51.27%、56.00%。其中普洱茶熟茶中、高剂量组小鼠负重游泳时间与空白对照组相比，

呈极显著差异（$P<0.01$）。这说明中、高剂量组的普洱茶熟茶能极显著延长小鼠的负重游泳时间。有文献报道，机体中血乳酸（BLA）、血尿素氮（BUN）的含量与抗疲劳效果成负相关，而乳酸脱氢酶（LDH）活力与抗疲劳效果成正相关。试验探讨了不同剂量的普洱茶熟茶对小鼠BLA、BUN含量和LDH活力的影响。与空白对照组相比，普洱茶熟茶各剂量组小鼠运动后BLA、BUN的含量均有降低趋势，而LDH均有增高趋势。其中中、高剂量组小鼠的BLA含量分别比空白对照组降低了22.1%、27.8%，均呈极显著差异（$P<0.01$）；低剂量组小鼠的BLA含量比空白对照组降低了17.9%（$P<0.05$）。高剂量组小鼠的BUN含量比空白对照组降低了17.1%（$P<0.01$）；中剂量组小鼠的BUN含量比空白对照组降低了12.6%，呈显著差异（$P<0.05$）。高剂量组小鼠的LDH活力比空白对照组增高了26.3%（$P<0.01$）；中剂量组小鼠的LDH活力比空白对照组增高了15.2%（$P<0.05$）。这表明普洱茶熟茶能降低小鼠运动后BLA、BUN的含量，增加LDH活力，且以高剂量的普洱茶熟茶效果最佳。有研究表明，运动后MG和LG的含量与抗疲劳效果成正相关。实验结果表明，与空白对照组相比，普洱茶熟茶各剂量组小鼠运动后MG、LG的含量均有增高趋势。其中低、中、高剂量组小鼠的MG含量分别比空白对照组增高了28.8%、42.5%、26.0%，均呈极显著差异（$P<0.01$）。而在LG方面，中、高剂量组小鼠的LG含量分别比空白对照组增高了22.0%、24.9%，均呈极显著差异（$P<0.01$）；普洱茶熟茶低剂量组小鼠的LG含量比空白对照组增高了16.7%（$P<0.05$）。这表明普洱茶熟茶能明显提高小鼠运动后MG和LG的含量。

普洱茶源于天然，有着悠久的历史和文化底蕴，在人们的日常生活中具有重要的地位，在保健功效方面有着独特的优势，与合成药相比，具有费用低廉，无毒副作用等优点，每日饮茶不仅可以消除疲劳，更能全面补充身体的其他营养成分。因此，深入研究普洱茶的抗疲劳作用具有重要意义。

有资料表明，BLA、BUN和LDH活力水平也是反映肌体的有氧代谢能力和疲劳程度的重要指标。张冬英等研究结果显示，运动后低、中、高剂量组小鼠的BLA、BUN水平均显著低于空白对照组，而运动后中、高剂量组小鼠的LDH活力水平均显著高于空白对照组，这说明普洱茶熟茶可能通过增强LDH活力，清除肌肉中过多的乳酸，从而减少运动中乳酸的生成，伴随着尿素氮生成的减少，机体对负荷的适应性提高，达到延缓疲劳产生的效果，其中以高剂量组的效果最明显。

第三节 普洱茶抗氧化及清除自由基功效

一、普洱茶的抗氧化机制

近年来，自由基与多种疾病的关系，已愈来愈被重视，自由基生物医学的发展使得探寻高效低毒的自由基清除剂——天然抗氧化剂成为生物化学和医药学的研究热点。21世纪现代农业的一个重要内容也是寻求和利用农产品的新的生物活性物质，其中，抗氧化活性的研究至关重要。

抗氧化作用被认为是茶叶保健抗癌最重要的机理。普洱茶是一种特殊的后发酵茶，在高温高湿的渥堆过程中，以黄酮类、茶多酚为主的多酚类成分在湿热和微生物作用下，发生微生物转化、酶促氧化、非酶促自动氧化以及降解、缩合等复杂的化学反应，形成了化学结构更为复杂的酚类成分。如普洱茶中没食子酸、茶褐素等活性成分的含量显著增高，大大增强了普洱茶在抗氧化方面的作用潜力。

目前的研究报道表明普洱茶抗氧化机制大致通过以下三个途径：

（1）抑制自由基的产生或直接清除自由基。Lin等人报道了普洱茶浸提物具有很强的清除羟自由基能力和抑制氧化氮自由基（nitric oxide radicals）生成的能力。Lin等人的研究表明普洱茶提取物可有效地在Fenton反应体系中发挥自由基清除作用，保护DNA超螺旋结构，防止链断裂。揭国良等人的研究表明普洱茶水提物中的乙酸乙酯萃取层组分和正丁醇萃取层组分对DPPH和羟自由基均有较强的清除能力。此外有研究报道普洱茶提取物在Fenton反应体系中自由基清除作用，抑制巨噬细胞中脂多糖诱导产生NO的能力与螯合铁离子作用均强于绿茶、红茶、乌龙茶提取物；200微克/毫升普洱茶水提物的抑制脂质过氧化能力与其他茶类（绿茶、红茶和乌龙茶）相比无显著性差异，但当浓度增至500微克/毫升时，普洱茶水提物抑制能力均强于其他茶类。

（2）抑制脂质过氧化。据Yang等人的研究表明，在动物实验中，普洱茶具有降低血清胆固醇水平的功效，但对血清中的高密度脂蛋白和甘油三脂的水平没有改变，高密度脂蛋白与总胆固醇的比值有显著提高，动脉粥样硬化指数得以降低。对于由高胆固醇饮食引起的肝脏重量增加，肝脏胆固醇含量提高和甘油三脂含量升高，普洱茶只能够降低肝脏的胆固醇含量，对肝脏的甘油三脂的含量没有明显降

低。在降低胆固醇方面，萧明熙等以普洱茶的水提物（PET）为试验材料，探讨其于体外试验中对胆固醇生物合成的影响，以及在活体动物中是否具有降血脂的效果，研究结果发现PET在人类肝癌细胞株（HepG2）模式系统中可以减少胆固醇的生物合成，且其抑制作用在生成甲羟戊酸之前。在动物实验中，也证实普洱茶有抑制胆固醇合成的效果，此外，还可降低血中的胆固醇、甘油三醋及游离脂肪酸水平，并增加粪便中胆固醇的排出。孙璐西等研究表明普洱茶水提物具有明显的抗氧化活性，清除自由基，降低低密度脂蛋白不饱和脂肪酸的含量，以降低低密度脂蛋白的氧化敏感度。

（3）螯合金属离子。Duh等人报道普洱茶水提物具有螯合金属离子，清除DPPH自由基和抑制巨噬细胞中脂多糖诱导产生NO的效果。普洱茶有很强的抗氧化性，能够清除DPPH自由基和抑制Cu^{2+}诱导的低密度脂蛋白（LDL）氧化。

二、普洱茶的抗氧化功能

东方利用普洱茶粉，对照选用浙江省龙游茗皇天然食品开发有限公司的绿茶粉与红茶粉浸提液开展动物实验。实验设对照组、绿茶组、红茶组与普洱茶组。采用灌胃给药小鼠，各茶组剂量为0.9克/（千克·天），对照组灌胃相当量的对照液（蒸馏水）。每天记录各组小鼠的体重变化。给药3周后断头取血与肝组织，进行MDA含量、SOD活性、GSH-PX活性和体外自由基（DPPH）测定。

东方首先测定了用于本研究中各茶粉的化学成分。其中绿茶中的多酚含量均高于红茶与普洱茶，而黄酮类含量则低于红茶与普洱茶。红茶与普洱茶具有相当量的没食子酸与咖啡碱，且含量均高于绿茶。在绿茶与普洱茶中仅检测到少量的TF3G，红茶中的茶黄素含量高于绿茶与普洱茶。绿茶的儿茶素以EGC与EGCG为主，约占总量的70%以上。在DPPH反应体系中，各茶粉浓度的对数与清除率存在着线性关系（$P<0.01$）。由线性方程得出的抑制50% DPPH时所需浓度（IC_{50}），结果表明各茶粉对DPPH自由基的清除效果由强到弱依次为绿茶>红茶>普洱茶。整个实验期，小鼠体重基本上没有变化，实验末期体重虽略有下降，但与实验初期比较并无显著性差异（$P>0.05$）。与对照组相比，PTW组能显著降低小鼠血清中MDA含量（$P<0.05$），绿茶组与红茶组均无显著性差异（$P>0.05$）。红茶组（$P<0.01$）与普洱茶组（$P<0.05$）均能降低小鼠肝组织中MDA的含量，绿茶组与对照组相比，无显著性差异（$P>0.05$）。绿茶组与BTW组能提高小鼠肝组织中的SOD活性，与对照组相比达到极显著差异（$P<0.01$），而PTW组则对小鼠肝组织中的SOD活性具有抑制作用，与对照组相比达到极显著差异（$P<0.01$）。在小鼠血清中，各组均未检测到SOD活性。绿茶组与红茶组均能显著提高小鼠血清中GSH-PX活性，与对照组相比达到极显著差异（$P<0.01$），普洱茶组对小鼠血清中GSH-PX活性无显著性影响（$P>0.05$）。肝组织中的GSH-PX活性结果表明，三类茶组均能提高小鼠肝脏中GSH-PX活性，与对照组相比，绿茶组与红茶组达到显著差异（$P<0.05$），普洱茶组则达到极显著差异（$P<0.01$）。

不同的发酵程度影响了绿茶、红茶与普洱茶的多酚组成与含量差异。普洱茶发酵过程中由于有微生物参与作用，在漫长的温、湿的环境条件下其多酚类的变化更为复杂，且具有一定量的黄酮类化合物。在本次研究中，红茶（水提物）仅含有约1%茶黄素，可能是大部分茶黄素进一步氧化转化为茶红素或茶褐素等物质、普洱茶中大多数儿茶素已被氧化，仅存在一定量的GC（约占水提物的5.4%），且高于绿茶与红茶。尽管普洱茶中多酚的含量比绿茶类少，但用超滤分离法得到的普洱水提物经分析后得高分子量物质（MW>3000 Daltons）多于50%（w/w），且普洱茶中没食子酸的含量高于绿茶。

研究表明茶叶中多酚类化合物清除自由基的能力已远远超过维生素C和维生素E等抗氧化剂。现有研究报道表明普洱茶提取物在Fenton反应体系中清除自由基作用，抑制巨噬细胞中脂多糖诱导产生NO的能力与螯合铁离子作用均强于绿茶、红茶、乌龙茶提取物。结果表明体外清除DPPH自由基能力大小依次为绿茶>红茶>普洱茶。

超氧化物歧化酶（SOD）与谷胱甘肽过氧化物酶（GSH-PX）是机体内清除自由基的重要抗氧化酶，对机体的氧化与抗氧化平衡起着至关重要的作用。本研究结果表明红茶与绿茶均能有效提高SOD活性，且红茶略高于绿茶，而普洱茶对SOD活性则起抑制作用，这与Kuo研究报道的基本一致。本实验中各组血清中均未检测到SOD活性，可能是饲料中的高脂成分进入血液后较易形成ROO·，RHOO·等类型的自由基，大量自由基转化后的下游产物抑制了SOD的活性。三类茶对GSH-PX的活性均有促进作用，且普洱茶对肝组织中的GSH-PX活性促进作用均强于绿茶与红茶。MDA是氧自由基攻击生物膜中的不饱和脂肪酸而形成的脂质过氧化物，可反映出机体内脂质过氧化和机体细胞受自由基攻击的损伤程度。研究报道表明当增至一定浓度时，普洱茶水提物抑制脂质过氧化能力强于绿茶与红茶，这可能解释了在本研究中与对照组相比普洱茶显著降低了MDA的含量，绿茶则无显著性差异。普洱茶一些特殊保健功能可能与存在的特异多酚类物质如儿茶素的寡聚体等密切相关。我们曾经报道了在普洱茶乙酸乙酯萃取层中分离出的ES层在清除轻自由基、超氧阴离子的能力及其对H_2O_2诱导HPF-1细胞损伤的保护作用均强于EGCG。这也表明了在普洱茶发酵过程中产生的一些未知的高分子量多酚类物质如儿茶素衍生物或聚合物可能具有与EGCG相当甚至更强的抗氧化功效。

普洱茶的化学成分非常复杂，多酚类、黄酮类、多糖类等化合物均具有较强的抗氧化活性。醋酸乙酯萃取部位为抗氧化活性部位，从该部位分离鉴定出的化合物主要有儿茶素类化合物、黄酮类化合物（山萘酚、槲皮素和杨梅素）以及黄酮的糖苷等，均具有较多的羟基及较强的自由基清除能力。没食子酸是普洱茶中的主要抗氧化活性成分之一。金裕范等比较云南5个产地普洱茶的抗氧化活性，选择存放3年的普洱饼茶，采用DPPH测定其抗氧化活性和自由基消除活性，研究结果表明5个产地的普洱茶提取物均具有一定的抗氧化活性，以云南大理下关产普洱茶的抗氧化能力最强，其EC_{50}值为8.88毫克/升，云南普洱思茅最弱，其EC_{50}值为21.81毫克/升，云南5个产地普洱茶抗氧化活性的强弱顺序依次为：大理下关普洱茶＞西双版纳普洱茶＞临沧普洱茶＞红河普洱茶＞思茅普洱茶。表明普洱茶可作为一种优良的天然抗氧化剂和自由基消除剂，云南不同产地普洱茶的抗氧化活性略有差异。

江新凤等采用高脂饲料饲喂法建立高脂血症大鼠模型，通过普洱生茶、熟茶、乌龙茶、药组分别灌胃，实验35天后，检测大鼠血液丙氨酸氨基转移酶

（ALT）、天冬氨酸氨基转移酶（AST）、总胆固醇（TCHO）、甘油三脂（TG）、高密度脂蛋白胆固醇（HDL-C）、低密度脂蛋白胆固醇（LDL-C）、谷胱甘肽过氧化物酶（GSH-Px）、微量丙二醛（MDA）、超氧化物歧化酶（SOD）的含量，以及观察大鼠的一般情况和肝、肾组织的病理变化，来观察普洱茶等对实验性高脂血症大鼠血脂水平调节和血管内皮细胞的保护作用。结果表明药、乌龙茶和普洱茶均能明显降低模型大鼠血液TCHO，TC，LDL-C和SOD含量，提高HDL-C，AST，MDA和GSH-PX的含量（$P<0.05$，$P<0.01$），其中，普洱茶作用显著优于药、乌龙茶。研究结论，药、乌龙茶和普洱茶均能显著调节机体的血脂水平，有效预防高脂血症和抗氧化等功能。

任洪涛等用同时蒸馏萃取法（SDE）富集普洱茶挥发性物质，并用GC-MS分析其化学组成，采用DPPH和FRAP法对不同发酵阶段普洱茶挥发性物质的抗氧化活性进行评价，分析抗氧化活性与主要成分含量的关系。结果表明，普洱茶在发酵过程中甲氧基苯类化合物的相对含量大幅增加；挥发性物质的DPPH自由基清除能力和FRAP总抗氧化能力随发酵程度的加深呈显著上升趋势，发酵出堆后分别提高了100%和296%；挥发性物质的DPPH自由基清除能力和FRAP总抗氧化能力与甲氧基苯类化合物和芳樟醇氧化物的相对含量具有显著正相关性。

陈浩比较分析了陈化时间分别为1年、3年、5年的普洱茶多糖中主要化学成分，评价了其体外抗氧化性能，同时研究了普洱茶多糖对四氧嘧啶诱导高血糖小鼠的餐后血糖、空腹血糖以及抗氧化状态的调节作用。研究结果表明，不同陈化时间的普洱茶多糖的含量和化学组成不同。5年陈普洱茶多糖（PTPS-5）得率最高（3.66%），3年陈普洱茶多糖（PTPS-3）次之（2.24%），1年陈普洱茶多糖（PTPS-1）得率最低（0.79%）。三种普洱茶多糖的蛋白质含量随陈化时间的增加而增加，PTPS-3和PTPS-5的糖醛酸含量也显著高于PTPS-1（$P<0.05$）。GC分析发现，尽管三种普洱茶多糖的单糖组成比例各不相同，但都是以半乳糖、阿拉伯糖、甘露糖为主，同时还有葡萄糖、鼠李糖、岩藻糖等单糖。分子量测定表明普洱茶陈化时间可以提升普洱茶多糖中低分子量多糖的含量。普洱茶多糖具有较强的抗氧化活性和突出的α-葡萄糖苷酶抑制能力，且和陈化时间有密切关系。在四种（ABTS自由基清除能力、DPPH自由基清除能力，FIC铁离子螯合能力、FRAP还原能力）不同的抗氧化评价体系下，PTPS-5具有最强的ABTS自由基清除能力（IC_{50}=0.49毫克/毫升），DPPH自由基清除能力（IC_{50}=1.45毫克/毫升），FRAP还原能力（浓度为1毫克/毫升时，FRAP值为1623.07，FIC亚铁离子螯合能力（（IC_{50}=0.73毫克/毫升）。此外，PTPS-5还具有最强的α-葡萄糖苷酶抑制能力（IC_{50}=0.063毫克/毫升），显著高于阳性对照阿卡波糖（IC_{50}=0.18毫克/毫升），而PTPS-3也具有和阿卡波糖相似的α-葡萄糖苷酶抑制能力（IC_{50}=0.19毫克/毫升）。不管是四种抗氧体系下的抗氧化活性还是对α-葡萄糖苷酶的抑制能力强弱都是PTPS-5>PTPS-3>PTPS-1。普洱茶多糖对四氧嘧啶诱导糖尿病小鼠体内抗氧化状态有积极调节作用。灌胃40毫克/千克剂量的PTPS-5能将小鼠血清以及肝脏组织中的MDA含量和SOD活性改善至和正常组小鼠无显著差异水平，GSH-Px活性甚至显著高于正常小鼠（$P<0.05$），说明普洱茶多糖对糖尿病小鼠体内的抗氧化状态有积极的调节作用。

第四章 普洱茶的饮用品鉴

第一节 普洱茶五大主要茶区

普洱茶茶山分布广泛，从大的区域看主要分布在澜沧江中下游流域。本书介绍了普洱茶的五大主要产区，对茶区内具有代表性的热门山头茶进行概述。

一、西双版纳茶区

（一）茶区地理位置和环境气候

西双版纳位于云南的最南端，地处北纬21°08′~22°36′，东经99°56′~101°50′，与老挝、缅甸山水相连，和泰国、越南近邻，土地面积近2万平方千米，国境线长达966千米。西双版纳茶区内主要包括勐腊茶区和勐海茶区。其中勐腊茶区所在地勐腊县地处云南省最南端，位于北纬21°09′~22°23′，东经101°05′~101°50′，地处北回归线以南。东部和南部与老挝接壤，西边与缅甸隔江相望，西北与景洪市相接，北面与普洱市江城县毗邻。国境线长达740.8千米（中老段677.8千米，中缅段63千米）。勐腊县城距省城昆明868千米，距州府景洪172千米。勐腊茶区海拔700~1900米，年平均气温17.2℃，年平均降水量1500~1900毫米。茶区内主要民族有基诺族、傣族、哈尼族、瑶族，饮食以酸、辣、生、腥为主。建筑风格大多以傣式建筑居多，近几年因普洱茶价格不断升高，茶农收入大幅增加，所以基本上每个村寨都盖上新式楼房。

勐海茶区的主要归属地勐海县位于云南省西南部、西双版纳傣族自治州西部，地处东经99°56′~100°41′、北纬21°28′~22°28′。东接景洪市，东北接普洱市，西北与澜沧县毗邻，西和南与缅甸接壤。国境线长146.6千米。东西最长横距77千米，南北最大纵距115千米，总面积5511平方千米，其中山区面积占93.45%，坝区面积占6.55%。县城勐海镇距省会昆明776千米，距州府景洪40千米。勐海县属热带、亚热带西南季风气候，冬无严寒、夏无酷暑，年温差小，日温差大，依海拔高低可分为北热带、南亚热带、中亚热带气候区。年平均气温18.7℃，年均日照2088小时，年均降雨量1341毫米，全年有霜期32天左右，雾多是勐海茶区的特点，平均每年雾日107.5~160.2天。

（二）茶区分布

西双版纳几乎全境产茶，景洪境内有古六大茶山之攸乐古茶山（现称基诺山）。景洪以东为勐腊茶区，是历史上有名的茶马古道源头，古六大茶山主要所在地。景洪以西为勐海茶区，是近年较热门新六大茶山主要所在地（除景迈属普洱澜沧茶区外，其他五山均在勐海境内）。因澜沧江穿景洪而过，所以大家又习惯将古六大茶山称为江内茶区，而将新六大茶山称为江外茶区。

1. 勐腊茶区

"勐腊"系傣语，"勐"意为平坝或地区，"腊"意为"茶""茶水"即"献茶水之地"。传说释迦牟尼巡游到此时，人们献很多茶水，喝不完的倒在河里，此河名"南腊"，即"茶水河"，"勐腊"因此得名。勐腊在西汉时属益州郡哀牢地，东汉属永昌郡鸠僚地，隋朝属濮部，唐（南诏）时属银生节度，宋代属景陇王国，宋淳熙七年（公元1180年），归属傣族首领帕雅真，元代属彻里路军民总管府，明清属车里宣慰使司，明隆庆四年（公元1570年），车里宣慰司将其辖区划分为十二版纳，勐腊县境内勐腊、勐伴为一版纳，勐捧、勐润、勐满为一版纳，整董、倚邦、易武为一版纳，清雍正七年（公元1729年）置勐腊土把总。1913年属普思沿边行政总局第五区（勐腊），第六区（易武）行政分局，1927年第五区改置镇越县，第六区改置象明县。1929年象明县并入镇越县，属普洱道。1949年11月6日解放，成立镇越县人民政府，隶属宁洱专区。1953年撤销镇越县，置版纳易武、版纳勐腊、版纳勐捧和易武瑶族自治区，属西双版纳自治区（州）。1957年并为易武、勐腊两个县级版纳。1958年并为易武县，1959年改名勐腊县。2002年3月22日，云南省人民政府批准：撤销勐腊县勐腊乡、勐腊镇，设立勐腊镇，镇政府驻原勐腊乡政府驻地曼列村，将原勐腊乡景飘行政村毛草山、桃子箐、纳秀3个村民小组划

归勐伴镇会落行政村管辖。2004年9月30日，云南省人民政府批准：撤销曼腊彝族瑶族乡、勐润哈尼族乡，原曼腊彝族瑶族乡管辖的行政区域划归易武乡管辖，原勐润哈尼族乡管辖的行政区域划归勐捧镇管辖。

勐腊境内主要产茶区集中在北部山区，历史上有名古六大茶山除攸乐外，全在勐腊境内的易武乡和象明乡，主要茶山分布如下：

易武：曼秀、三丘田、落水洞、麻黑、丁家寨、刮风寨、老街、弯弓。

蛮砖：曼庄、曼林、曼迁。

革登：新洒坊、值蚌、新发。

莽枝：江西湾、秧林、董家寨、红土坡。

倚邦：曼松、曼拱、架布、河边、麻栗树。

攸乐：亚诺、龙帕、司土老寨、么卓。

整个勐腊茶区内茶种较杂，有野生型古树、过渡型古树（变异紫径茶），但主要以大叶种茶为主，易武大叶茶相比勐海大叶茶而言，叶片更大，更细长，不显毫，茶区内还特有当地称为柳条茶的小叶种茶，而小叶种茶又以曼松茶名气最大。

2. 勐海茶区

勐海茶区主要在勐海县。勐海县可分为5个气候区：

北热带。为海拔低于750米的打洛、勐板、布朗山的南桔河两岸河谷地区及勐往的勐往河和澜沧江两岸河谷地区。

南亚热带暖夏暖冬区。为海拔750～1000米的勐满、勐往坝区布朗山南桔河两岸。

南亚热带暖夏凉冬区。为海拔1000~1200米的勐海、勐遮、勐混、勐阿（包括纳京、纳丙）、勐往的糯东。

南亚热带凉夏暖冬区。为海拔1200~1500米的勐阿的贺建，勐往的坝散，勐宋的曼迈、曼方、曼金，格朗和的黑龙潭、南糯山，西定的曼马、南弄，巴达的新曼俄、曼皮、曼迈、章朗和勐冈全境。

中亚热带区。为海拔1500~2000米的西定、巴达、格朗和、勐宋4个乡的大部分地区及勐满的东南至东北面。

勐海县境内居住着傣族、哈尼族、拉祜族、布朗族、汉族、彝族、回族、佤族、白族、苗族、壮族、景颇族等25个民族，其中傣族、哈尼族、拉祜族、布朗族是本地的四大主体民族。主要茶山分布如下：

北部：勐海勐宋、滑竹梁子、那卡、保塘。

西部：南峤、巴达、章朗、曼糯、贺松。

东部：帕沙、南糯（多依、半坡、南拉、姑娘、石头、拔玛）、曼迈、贺开（邦盆、广别、曼弄、广冈）。

南部：景洪勐宋、布朗山（老班章、新班章、老曼娥、坝卡囡、坝卡竜、吉良）。

勐海茶区内茶种主要以勐海大叶种为主（勐海大叶种发源于南糯山），勐海大叶种茶芽头肥硕，叶片肥厚，毫浓密，茶梗粗壮。

勐海茶区传统家家自采制茶，且采摘规范，通常采摘标准以一芽二叶居多，茶区内早期很少有大规模的初制加工所，但近几年很多实力生产厂家在茶区内或承包

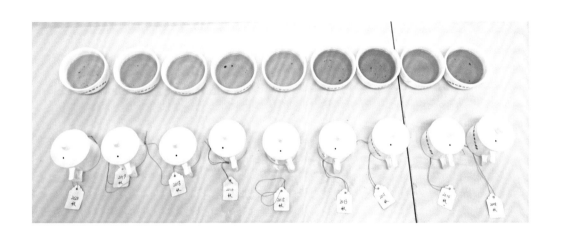

合作，或收购鲜叶加工，初制较为发达，茶叶外形松紧适度、芽肥毫密，色泽油润光亮，新茶汤色金黄明亮，苦味较重、叶底匀亮富韧性，陈放后则香气较好，汤质较厚、回甘快且持久。

二、普洱茶区

（一）茶区地理位置和环境气候

普洱市位于云南省西南部，地处北纬22°02′~24°50′、东经99°09′~102°19′，东临红河、玉溪，南接西双版纳，西北连临沧，北靠大理、楚雄。东南与越南、老挝接壤，西南与缅甸毗邻，国境线长约486千米（与缅甸接壤303千米，老挝116千米，越南67千米）。普洱市南北纵距208.5千米，东西横距北部55千米、南部299千米，总面积45385平方千米，是云南省面积最大的州（市）。市级机关驻思茅区的思茅镇，海拔1302米，距省会昆明415千米、空中航线305千米，乘飞机35分钟可抵达。茶区内主要民族有布朗族、瑶族、哈尼族、彝族、傣族、佤族、拉祜族等，民族风情迥异多彩。

普洱由于受亚热带季风气候的影响，大部分地区常年无霜，冬无严寒，夏无酷暑，享有"绿海明珠""天然氧吧"之美誉。普洱市海拔在317~3370米之间，中心城区海拔1302米，年均气温15~20.3℃，年无霜期在315天以上，年降雨量1100~2780毫米，负氧离子含量在七级以上。

（二）茶区历史与茶山分布

清道光《普洱府志》"六茶山遗器"载，早在1700多年前的三国时期，普洱府境内就已种茶，而最早在历史文献中记载普洱茶种植的人，是唐代咸通三年（公元862年）亲自到过云南南诏地的唐吏樊绰，他在其著《蛮书》卷七中云："茶出银生城界诸山，散收无采造法。蒙舍蛮以椒姜桂和烹而饮之。"银生城即今普洱地区的景东县城，景东城即是唐南诏时的银生节度所在地，银生节度辖今思茅地区和西双版纳州，历史记载说明，早在1100多年前，属南诏"银生城界诸山"的思普区境内，已盛产茶叶。明代万历年间的学者谢肇淛在其著《滇略》中，已提到"普茶"（即普洱茶）这个词，该书曰："士庶所用，皆普茶也，蒸而成团"。清光绪二十三年（公元1879年）以后，法国、英国先后在思茅设立海关，增加了普洱茶的出口远销，普洱茶马古道随之兴旺。作为文物遗迹的，今还有思茅三家村社坡脚寨茶马古道、卡房高酒房茶马古道、普洱那柯里茶马古道、茶庵塘茶马古道及景谷、镇沅、景东、墨江茶马古道、古驿站，石上马蹄印，记录下了当年运茶马帮的历史。普洱茶区主要的分布地在：

宁洱：困鹿山、新寨、板山。

澜沧：景迈、芒景、邦崴、帕赛。

江城：田房。

景谷：秧塔、苦竹、文顶山、黄草坝、龙塘、团结。

墨江：迷帝、景星。

镇沅：千家寨、老乌山、田坝、马邓、勐大、振太。

景东：御笔、金鼎、老仓福德、漫湾。

普洱茶区作为普洱茶的传统产茶大区域，具有较高的历史地位，无量山横穿整个茶区，茶区内茶种繁多，镇沅千家寨的野生茶，大、中、小叶混生的困鹿山，出产大白茶的景谷秧塔，大面积藤条茶群居的老乌山等。而后期思茅、宁洱、西盟等地则大面积种植了一些新的名优茶种，比如：云抗十号、雪芽100号、紫芽、紫娟等，丰富了适制普洱茶的品种。普洱茶区内所产普洱茶多以香柔为主，比如无量山茶的甜水、景迈茶特有的兰香等。

（三）茶叶特点

茶区内传统晒青毛茶以混采、高温快杀青、紧揉、日光暴晒的工艺为主，故传统工艺茶条黑紧，涩感强，多用于拼配或熟茶的渥堆发酵，而近几年因普洱山头茶一路受捧，产茶区域越来越细化，工艺也有较大改进，思茅茶区、景谷茶区等制茶工艺与勐海茶区接近。茶区内最为热门的景迈茶大多乔木混生于雨林之中，制作出的晒青毛茶外形匀称显毫，光鲜度好，开泡香气高扬，茶汤黄亮，涩度较低，叶底黄润。

三、临沧茶区

（一）茶区地理位置和环境气候

临沧市位于云南省的西南部，地处东经98°40′~100°34′、北纬23°05′~25°02′，临沧市东部与普洱市相连，临沧市西部与保山市相邻，临沧市北部与大理白族自治州相接，临沧市南部与邻国缅甸接壤。临沧市地势中间高，四周低，并由东北向西南逐渐倾斜。境内最高点为海拔3429米的永德大雪山，最低点为海拔450米的孟定清水河，相对高差达2979米。

临沧市属于横断山脉南延部分，是滇西纵谷区，境内老别山、邦马山两大山脉，永德大雪山、临翔大雪山和双江大雪山构成山脉主峰。澜沧江、怒江为临沧两大水系，两江境内有罗闸河、小黑江、南汀河、南棒河、永康河等。临沧属于亚热带低纬度山地季风气候，四季温差不大，干湿季节分明，对植物生长和茶树繁殖有着极其有利的一面。

临沧市辖临翔区、凤庆县、云县、永德县、镇康县和双江拉祜族佤族布朗族傣族自治县、耿马傣族佤族自治县、沧源佤族自治县8个县（区），77个乡（镇、街道），947个行政村（社区）。总面积2.45万平方千米。

（二）茶区种质资源与茶山分布

在云南省4个主要产茶区中，临沧市面积居首，采摘面积近90万亩，占全省面积的近三分之一，居云南全省茶叶产量之首。临沧主要有23个民族，其中史前的百濮（佤族、布朗族、德昂族）都是与云茶息息相关的民族。在临沧境内，有大面积野生茶树群落和栽培型古茶园。

临沧茶树资源丰富，经研究部门调查，全市共4个茶系，8个品种。全市7县一区都有丰富的野生茶树群落和栽培型古茶园。

（1）野生茶：最具有代表的就是双江勐库野生茶树群落、沧源糯良贺岭村大黑山野生茶树群落、耿马芒洪大浪坝野生茶树群落。

（2）栽培型古茶园：临沧市有百年以上树龄的栽培型古茶园约有23160亩，其中凤庆10600亩、云县5460亩、临翔区5000亩、双江县2000亩、沧源县100亩、耿马、永德、镇康等地均有分布。

临沧茶区主要分布如下：

临翔区：邦东、那罕、昔归。

双江：坡脚、邦骂、小户赛、坝歪、糯伍、南迫、冰岛、老寨、地界、坝卡、邦丙、邦改、大户赛、邦木、亥公。

镇康/永德：永德大雪山、鸣凤山、忙波、忙肺、马鞍山、岩子头、汉家寨、勐捧、梅子箐。

耿马/沧源：户喃、帕迫。

云县：白莺山、核桃岭、大寨。

凤庆：香竹箐、平河、岔河、永新。

四、保山茶区

（一）茶区地理位置和环境气候

保山茶区地处云南省西部，地势北高南低向南贯穿，澜沧江通过东部。在云南四个主要产茶区中，纬度最高、平均海拔最高、气温最低、雨量最少。辖区保山市、昌宁县、腾冲市、龙陵县、施甸县等地，都有大面积的茶叶生产。保山境内自然条件优越，适宜茶树生长，茶树品种资源丰富，是云南"滇红"及普洱茶的重要产地，1986—1987年，昌宁、腾冲、龙陵3县（市）被列为全国首批优秀茶基地县和国家出口红茶商品基地县。全市有10万亩无性系良种茶叶基地，15万亩无公害茶叶生产基地。

（二）茶区种质资源与茶山分布

昌宁县位于澜沧江中下游米明山秀水之间，境内气候十里不同天，海拔相对高度较大，形成了低热、温热、温凉、高寒的立体气候。森林覆盖率达46.7%。昌宁县境内的茶树品种资源十分丰富，茶园主要有箐茶和包洪茶。箐茶于1981年发现在街水炉阿甘梁子原始森林内，一般树高700～800厘米。树幅300厘米×500厘米，胸径8～10厘米，叶形椭圆，叶色绿，叶质薄软，叶面微微隆起，芽有多层鳞包裹，叶面积71平方厘米。包洪茶属于大理茶种类，在大田坝乡和漭水镇相联界的狮子塘梁子原始森林中，分布的野生古树茶资源，其中最大的一株高1000厘米，树幅600厘米×700厘米，胸径15厘米。栽培型古树茶、茶园主要品种有大理茶、腾子茶和昌宁大茶。

龙陵县古树茶资源，其中一株树高18.2米，树幅5.8米，干径123厘米，树型乔木，树姿直立，叶片长宽13.3厘米×6.6厘米，叶椭圆形，叶

厚有光泽，叶色深绿。芽叶茸毛稀少。花大，平均花径5.8厘米，花瓣11片，子房有毛，柱头5裂。

腾冲县古树茶资源，其中一株树高7.7米，树幅2.5米，干径29.3米，数型乔木，树资直立，叶片长宽17.5厘米×6.6厘米。主要形态特征：叶椭圆形、叶面光滑、叶质厚，叶色深绿。芽毛稀少，芽色微紫，平均花径4.8厘米。子房有毛，柱头5裂。

茶区主要茶山简介：

（1）潞水镇潞水村黄家寨栽培型古茶树群。此地海拔在1840米，古茶树群分布面积达100亩，其中较古老茶树400多株相对集中，树龄在500年以上。

（2）潞水镇沿江村茶山河保家洼子野生古茶树。此地海拔2348米，有野生"红裤茶""报洪茶"群。

（3）潞水镇沿江村羊圈坡野生古茶树群。此地海拔2340米，古茶树群分布较集中，100多米地坎上有基部直径40厘米以上大茶树20株，株距3米左右，属大理茶。

（4）温泉乡联席村芭蕉林野生古茶树群。此地古茶树群分布面积较大，茶树基部直径60厘米以上的大茶树有1000株以上，其中最大茶树高15米，基部茎围2.85米，树幅6米×6米，属于大理茶。

（5）石佛山古茶树群。田园镇新华村石佛山海拔2140米，古茶树群分布面积较大，其中有较大古茶树5株，最大一株茶树当地俗称"柳叶青"，基部茎围3.03米，树高14.8米，树幅6米×8.4米。1997年7月，西南农业大学刘勤晋教授到实地考证该茶树属大理茶亚系栽培型古茶树，树龄在1000年以上。

（6）温泉乡联席村破石头栽培型古茶树。此地属栽培型古茶树群，其中最大一株高5.8米，树幅5.1米×5.4米，基部茎围2.6米，基部有四个分枝，小乔木批张型，叶色绿，茸毛多。属普洱茶，当地人称原（袁）头茶，是云南作为茶树原产地的见证之一。据国家"茶树种质资源系统鉴定评价"研究结果，该茶树茶多酚、儿茶素含量高，制茶品质优，香气高。

五、文山茶区

文山茶区地处滇东南文山州，北回归线横贯全境，地形西北高东南低，最低海拔106米，最高海拔2991米，大部分地区海拔1000~1800米，是海洋暖湿气流北上云贵高原的第一个迎锋面，气候温暖湿润，极适宜茶叶生长。茶叶栽培种植历史悠久，史料记载1716年开化府副总兵向康熙进贡"普洱茶四十圆，女儿茶八篓……"是目前已查证史料中云南最早的贡茶记录；茶树种质资源丰富，是全世界茶树种类分布最丰富的地区之一。当地所产茶叶水浸出物丰富，氨基酸含量较高，茶品具有"花香馥郁，鲜甜醇爽"的特点。

经2013年至2017年全州古茶树种质资源普查，文山州8县（市）均有茶树种质资源分布，且种类丰富、多样，仅初步调查并命名的就有广南茶、广西茶、大厂茶、五柱茶、厚轴茶、光萼厚轴茶、马关茶、皱叶茶、秃房茶、茶、普洱茶和白毛茶共9个种3个变种，约占全世界茶种种类的35%，且尚有未确定种属的资源，是全世界茶种植物分布最多的地区。西畴县为厚轴茶模式标本产地，以县名命名的野生茶种有"马关茶""广南茶"。据不完全统计，全州野生茶树种质资源共计24个茶树种质资源居群，分布总面积约12000公顷，株数约46万株。据不完全统计全州栽培型古茶树（园）面积3460余公顷，品种以当地原生白毛茶群体种为主，兼有少量普洱茶种；栽培型茶树种质资源主要分布在麻栗坡、西畴、马关、广南，4个县分布总面积约2500公顷，株数约57万株，最具代表性的古茶山有麻栗坡老山古茶山、西畴坪寨古茶山、马关古林箐古茶山、文山老君山古茶山、广南九龙山古茶山等古茶山。

第二节 普洱茶的贮存

普洱茶作为一种典型的后发酵茶,具有"越陈越香"的特点,普洱茶特别是普洱生茶被誉为活的有机体,它的感官品质和独特风味必须经历相当长时间的后发酵过程才能形成。

从感官品质来看,普洱生茶随着贮藏时间的延长,茶叶颜色由浅墨绿向微棕褐变化,滋味由苦涩转变为入口涩微苦回甘、滑顺,汤色由淡黄色转变为红黄色或琥珀色。熟茶的汤色由红褐色向褐红色变化,香气由清纯转为陈香持久,滋味转向醇厚甘甜。从普洱茶的化学成分组成含量和功能性来看,随着贮存时间的延长,不管是生茶

陈化中的普洱生茶

还是熟茶,其含有的茶多酚、儿茶素、游离氨基酸、茶红素、茶黄素含量、可溶性糖保留量都有明显的降低,黄酮类化合物增加、茶褐素大量积累,普洱茶抗氧化性、清除NO_2能力未呈现明显增强规律,而对α-淀粉酶抑制作用有所增加。

普洱茶陈茶的品质形成有两个重要的因素:一是原料的选择,只有选择用料精良,品质稳定的茶叶,才具有陈化的基础;二是贮存环境的选择,茶叶在贮存期间由于受水分、氧气、温度、光照、微生物等外界因素的影响,内含物会发生氧化、聚合等,使得普洱茶在色泽、滋味、香气等方面都发生变化,其中影响贮藏过程中普洱茶品质形成的存储条件主要包括:

一、温度

普洱茶放置的温度最好是常年保持在20~30℃之间(以25℃±3℃最宜),普洱茶的陈化过程是一个渐进的酶促反应,低于20℃,普洱茶固有酶的活性降低,但超过50℃,酶蛋白会出现变性,酶促反应基本停止,这都会减缓普洱茶的变化过程。同时温度对普洱茶陈香香气的形成有影响,鲍晓华的研究表明,普洱茶香气在低温下陈香明显,高温下陈香减退,太高的温度会使茶叶氧化加速,有效物质减少,影响普洱茶的品质。

二、相对湿度

贮藏室相对湿度应控制在65%左右，湿度增加可以促进茶叶转化，但湿度超过70%后，空气湿度会将茶叶释放的香味大量吸收，加速普洱茶香味释放，而超过80%后，茶品霉菌快速生长，容易使普洱茶有劣变与熟化现象产生，会出现辛辣的香气和滋味，汤色会出现浑浊现象。

三、其他贮存条件的要求

茶叶容易吸收杂气杂味，所以普洱茶存储要求不能有异味，一般要有专门的贮藏室。光照会分解茶的有效成分，紫外线会直接影响酶的活性和引起光化作用，因此普洱茶要选择避光保存。贮藏环境的适度通风透气也是一个影响陈化的重要因素，首先，偶尔的通风可以将茶内的陈宿杂味吹散，其次透气的环境可以增加茶叶和氧气的接触程度，有利于茶叶中微生物的繁衍，从而加速普洱茶的变化过程。

总之，在普洱茶的后发酵或陈化过程中，湿度的控制至关重要，不论是高温高湿还是低温高湿，都容易使茶叶发生如右图所示的霉变，而高温低湿和低温低湿都会导致茶叶陈化过程变得缓慢。在自然陈化的过程中，由于受到地域、气候条件变化的影响，适时地采取一些辅助方法改善温湿度，如南方夏季可用除湿机降低湿度，北方冬季可用暖气和加湿来营造合适的存储环境，提高普洱茶陈化的品质。

霉变的普洱茶

第三节 普洱茶的品鉴方法

一、普洱茶冲泡技艺

泡茶是指用开水将茶的内含物质浸出的过程。中国自古以来就很讲究茶的冲泡技术，累积了丰富的经验。泡茶的过程需要讲究茶叶、茶具、用水、环境、茶者冲泡技艺等的协调，才能扬长避短，彰显茶性。

中国有六大茶类，每一类茶从用料到制作工艺都有着各自的特性，要彰显出每个茶的最佳状态，所采用的冲泡方法也不尽相同。普洱茶从制作工艺的不同，分为生茶与熟茶；从形态上又有饼茶、砖茶、沱茶、散茶等之分；从用料上也有不同等级、不同季节、不同区域等的特点；还有新茶、中老期茶、存放区域不同等等很多的要素，造就了普洱茶多姿的个性。所以，在本节中根据普洱茶不同的特性，来详细介绍不同的冲泡技巧。

（一）品茗环境

1. 品茗环境设置要求

泡茶品茶是交际、放松、享受、思考的过程与媒介，对于品茗的环境有严格要求，如若人们身处过于嘈杂或杂乱不整洁的场所，对于品茶的心境有很大的影响，不仅无法用心去体味茶之美，反而还糟蹋了茶。所以，品茗环境的设置是泡茶品茶的必要前提。

品茗环境并不要求要奢华、富丽堂皇，反之，需要的是整洁、安静、高雅即可。在这样的环境下，才能静心下来，品味茶之韵，感悟茶道的精髓。

2. 品茗人文环境要求

品茶的环境除了硬件设施的要求外，人文环境也要与其相得益彰。首先，茶者着装打扮要素雅简洁，不可佩戴过多的装饰品或涂抹香气高扬的化妆品，不化浓妆，否则会喧宾夺主，甚至与品茶空间不相宜。其次，茶者举止要优雅，坐姿、站姿、行姿等要讲究仪态之美。最后，茶者用语要规范，不能使用不雅的语句，讲话要注意音量与语调，尽量不大声喧哗，以免影响茶室的雅致氛围。

（二）普洱茶冲泡器具

1. 茶具介绍

茶壶：用来泡茶的主器具。主要以陶土、瓷质、玻璃、金属材质为主。

茶壶

盖碗：又称"三才杯"，盖为天、托为地、碗为人，暗含天地人和之意。

盖碗

公道杯：均匀茶汤，分茶的用具，起到均匀茶汤的作用。

公道杯

滤网：过滤茶渣的用具。

滤网

品茗杯：品茗用的小杯。
杯托：承载品茗杯的器具。

品茗杯和杯托

茶巾：用于擦干壶底、杯底、茶台等的剩余之水。

茶巾

茶道六君子（茶道组合）：茶则、茶匙、茶针、茶漏、茶夹、茶筒。

茶道组合

茶荷：盛放干茶，赏茶之用。

茶荷

烧水具：煮水用具。现常用的有随手泡或铁壶、铜壶、陶壶等配合电磁炉、酒精灯等使用。

烧水具

茶刀（茶针）：用于撬取紧压茶的工具。

茶针

解茶盘：撬茶时，将紧压茶放置于解茶盘中，可以让撬好的茶叶集于其中，不会散落四处，也可保护桌面不受茶针的损伤。

解茶盘

壶承：放置于茶壶下方，有容纳水功能的器具。干泡法时常用器具。

壶承

水盂或水洗：用于盛废水的容器。干泡法时常用器具。

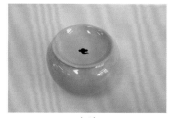

水洗

2.茶具的选择

茶具的种类繁多，分类标准不一，在泡不同茶叶时，选择不同材质的器具，所呈现出的香气、滋味也有很大差别，下面主要按不同材质器具的特点来分别介绍。

（1）陶土茶具

陶器中首推的应属宜兴紫砂茶具，紫砂茶具特指宜兴丁蜀镇所用的紫砂泥坯烧制后所制而成的。紫砂壶和一般陶器不同，其里外都不施釉，采用当地的紫泥、红泥、团山泥，用手工拍打成形后焙烧而成。紫砂壶的烧制温度在1100～1200℃之间，属高温烧成。

紫砂壶具始于宋代，至明清时期达到鼎盛，并流传至今。紫砂壶是集诗词、绘画、雕刻、手工制造于一体的陶土工艺品，造型美观，风格多样，不仅具有极高的收藏价值，还是泡茶贮茶的佳具，有"泡茶不走味，贮茶不变色，盛暑不易馊"的美名。

紫砂器具的优点：

① 紫砂泥质是双重气孔结构，气孔微小，并且密度很高。紫砂壶泡茶既不夺茶的香气，又不会造成茶水有熟汤的味道。

② 紫砂的透气性极佳，用紫砂罐储存普洱茶，不仅透气性好，而且避光阴凉不潮湿，对普洱茶的后期转化非常有利。

③ 紫砂壶能很好的吸取茶味。紫砂壶经过久用之后，其内壁会堆积茶垢，经过长年养护的紫砂壶，在空壶中注入沸水，也会闻到茶的香气。

④ 紫砂壶耐极冷极热性能好，不会因温度突变而胀裂。紫砂泥属砂质陶土，传热慢，不易烫手，还可在火上进行加温。并且紫砂壶的保温时间较长，用来泡中老期茶是再好不过的佳具了。

（2）瓷质茶具

瓷器是中国文明的一种象征，瓷器茶具又可分为白瓷茶具、青瓷茶具、黑瓷茶具、彩瓷茶具等。

瓷器由于有一定的吸水率，且导热系数中等，并且泡茶时在水流的冲击下，可让茶叶和杯壁产生碰撞，能很好地激发出茶叶的芳香物质，所以用瓷器泡茶时茶的香气会比较明显且持久一些。且瓷器的导热没有紫砂那么慢，瓷器内温度也没有紫砂高，所以用瓷质的茶具来冲泡较细嫩的茶叶，不仅不会有熟汤感，还可以很好地彰显出茶叶的清香。用瓷质的品茗杯来品茶时，不仅不易烫口，还能很好地欣赏茶汤汤色之美。

（3）玻璃茶具

在中国古代，玻璃被称为"琉璃"，中国的琉璃制作技术虽起步较早，但直到唐代，随着中外文化交流的增多，西方琉璃器具的不断传入，中国才开始烧制琉璃茶具。

玻璃茶具导热性能好，并且质地透明，一般在泡细嫩的绿茶时用得比较广泛，不仅可以很好的观赏茶叶在水中舒展的姿态，还可以很直观地欣赏到茶汤。由于普洱茶冲泡时要求要较高的水温，所以不建议使用玻璃壶进行冲泡。但可以选用玻璃材质的滤网和品茗杯，因为玻璃材质不具有吸附性，不会影响茶味。使用玻璃品茗杯还可以很好地欣赏茶汤色泽，也能快速降低茶汤温度，不易烫口。

3.茶具选用效果对比

各类茶具使用的效果用表4-1来详细说明

表4-1　茶具使用效果对比表

茶具分（按材质分）	导热性	吸附性（吸水、吸味性）	茶汤香气	茶汤口感	易泡的普洱茶类	不易泡的普洱茶类
紫砂茶具	低	高	一般	很饱满	普洱熟茶；具有一定年份的普洱生茶	等级较高、年份较新细嫩的普洱生茶
瓷质茶具	中	中	较好	较饱满	大部分普洱茶类都可选用	5年以上的中老期生、熟茶
玻璃茶具	高	低	略差	一般	等级特高，细嫩的新茶或散茶	大部分普洱茶冲泡尽量不选用玻璃茶具

（三）普洱茶撬茶方法

1.紧压茶的撬茶方法：

（1）将要撬取的紧压茶放置于茶盘中。

（2）打开包装纸，将茶饼底部有凹心的一面朝上。

（3）用包装纸遮盖住茶饼至一半的位置，避免手直接接触茶叶，左手扶按住凹心后边缘，右手持茶针，左右手要保持平行线，茶针不可与左手相对，更不能朝向自己。

（4）从茶饼中间凹心处向前方向插入茶针，用巧劲撬下茶块。

紧压茶的撬取

2. 注意事项

在撬茶的过程中要注意，不要用手直接接触茶饼，避免手上的细菌等杂物沾到茶叶上。

持茶针的手方向和扶茶饼的手方向要一致，保持平行，不可相向而对，以避免茶针戳伤手。

由于普洱茶在压制时，一饼茶会分为撒面、盖茶、心茶几层，每层用料不一，只有部分茶压制时是会从内而外都用相同的料。所以在取茶时，若只撬取了一个单层的茶叶，在品茗时就无法充分地感受这饼茶的综合风味。

在解茶时要注意取茶的整碎度，如果茶块撬的过大，不易浸泡出滋味；如撬茶撬的太碎，内含物浸出太快，也会影响茶汤滋味。

（四）泡茶用水

1. 择水

"茶性必发于水。八分之茶，遇十分之水，茶亦十分矣；八分之水，试十分之茶，茶只八分耳。"《梅花草堂笔谈》中这样记载着水与茶的关系。还有陆羽的《茶经·五之煮》中也对泡茶用水做了详细的研究与记载："其水用山水上、江水中、井水下。"可见，水为茶之母，水能载茶，亦能覆茶。水质直接影响茶质，泡茶水质的好坏会影响到茶的色、香、味的优劣。

（1）鉴水五要素

清——水质洁净透彻。

活——有源头而常年流动的水，在活水中细菌不易大量繁殖。

轻——分量轻，比重较轻的水中所溶解的钙、镁、铁等矿物质较少。

甘——水含在口中给人的甜美感觉，不能有咸味或苦味。

冽——水在口使人有清凉感。

（2）茶汤与水的酸碱度关系

汤色对pH值的高低很敏感，当pH值小于5时，对汤色影响较小；如超过5，总的色泽就相应地加深，当pH值达到7时，茶黄素倾向于自动氧化而损失，茶红素则由于自动氧化而使汤色发暗，以致失去汤味的新鲜度。因此，泡茶用水的pH值不宜超过7，宜用中性水或弱酸性水，否则将降低茶汤的品质。

（3）茶汤与软硬水的关系

泡茶用水，通常分软水和硬水两种。1升水中钙、镁离子含量低于8毫克的为软水,超过8毫克的称为硬水。

用硬水泡茶,会影响茶汤滋味,不仅口感平淡,香气也不高。所以宜用软水泡茶。

2. 煮水

普洱茶需要用沸水来进行冲泡才能使茶叶充分舒展,香气、滋味得到最好的呈现。虽然当下也出现了冷泡法,即茶叶用冷水经过十多个小时的浸泡泡出滋味的方法,但这种方法还比较少见,而且也不适用于紧压的普洱茶类,所以在冲泡普洱茶之前要学会煮水的方法。

对于煮水的讲究,早在古代就有记载。陆羽的《茶经·五之煮》中指出:"其沸,如鱼目,微有声,为一沸;缘边如涌泉连珠,为二沸;腾波鼓浪,为三沸。已上水老不可食也。"明许次纾《茶疏》中也进一步提到:"水,入铫,做须急煮,候有松声,即去盖,以消息其老嫩,蟹眼之后,水有微涛,是为当时,大涛鼎沸,旋至无声,是为适时,过则汤老而香散,决不堪用。"

除了文献的记载,现代科学也对此有研究。煮水过程中水中的矿物质离子会产生变化。水中的钙、镁离子在煮沸过程中会沉淀,煮水时间过短,钙、镁离子尚未沉淀完全,会影响茶汤滋味。久沸的"老水",水中含有微量的硝酸盐在高温下会被还原成亚硝酸盐,这样的水不利于泡茶,更不利于人体健康。

3. 水温

泡茶的水温高低是影响茶叶水溶性内含物浸出和香气挥发的重要因素。根据不同茶叶特性,要掌握不同的泡茶水温。普洱茶冲泡的水温一般在90~100℃之间,如若沸点为100℃,这里所指的沸点以下的温度是在水煮沸后,通过置凉或人为使其变凉,达到所需的温度,而不是将未煮沸的水直接用来进行冲泡。

由于普洱茶用料级别不同,年份不同,形制不同,所以在冲泡时对水温的要求也各不相同。若水温掌握不当,则滋味也会受到很大的影响。例如,冲泡老茶时用较低的水温,则茶中的物质不能够被充分浸出,香气和茶汤饱满度都得不到充分展现;若用太高温度的水来冲泡用料级别较高的茶,不仅会影响茶的鲜爽度,造成苦涩味过度,甚至还会出现茶叶闷熟的滋味。

所以在冲泡前,要充分掌握茶品的特性,调节泡茶水温。一般而言,冲泡时,紧压状普洱茶要比散状普洱茶用水温度高;用料等级低的普洱茶要比用料等级高的用水温度高;年份久的普洱茶要比年份短的用水温度高;用料、形制、年份等统一的条件下，普洱熟茶要比普洱茶的晒青毛茶的用水温度略高。

4. 注水

在冲泡茶叶时，注水的技巧也会直接影响到茶汤的质量。在注水时，首先要求水流要平稳，不能过缓过急；其次，水流不能直接冲淋到茶身，要沿壶壁缓缓注入。普洱茶因为形制是紧压的特殊性，有醒茶的步骤，在醒茶时，注水水流可均匀稍快，让茶块可以稍微翻腾，与水充分的接触，唤醒茶性；当泡到3~4泡之后，这时的茶叶已充分舒展开来，注水要匀缓，让茶叶内含物质自然浸出，若这时水流过快，不仅会使汤色变浑，还影响茶汤滋味；当茶叶泡至味寡淡时，可以采用高温急速注水的方式，提高叶底温度，加快内含物的浸出。

还需要注意的是，当泡散茶或较碎的茶时，由于茶叶内含物浸出会较快，无论醒茶还是正式冲泡，都要保持匀缓的水流，避免让茶叶在壶内产生较大的回旋转动，若水流过急，会让茶汤变得浑浊，还会让茶汤滋味难以掌控。

（五）茶水比与冲泡频次

茶水比指的是泡茶用水与茶叶的比例。根据场合、人数的不同，器具规格、冲泡茶叶的不同选择，茶水比也稍有不同。表4-2根据不同普洱茶的不同特性，来掌握茶水比与冲泡频次的区别。

表4-2 各类普洱茶冲泡的茶水比

普洱茶类型	器具	容量	投茶量	冲泡频次
茶菁细嫩的紧压茶	盖碗	150毫升	6~7克	12~13泡
茶菁粗壮的紧压茶	盖碗或紫砂壶	150毫升	7~8克	14~15泡
茶菁粗老的紧压茶	紫砂壶	150毫升	7~8克	14~15泡
散茶或细碎茶叶	盖碗	150毫升	6~7克	10~12泡
5~10年中期茶	紫砂壶	150毫升	8~10克	15~20泡
10年以上老茶	紫砂壶	150毫升	10克以上	20泡以上

表4-2所标示数据仅为建议茶水比与冲泡频次，根据人数不同、茶具容量不同、茶叶特性的区别，也可在此基础上稍作增减。

（六）行茶礼仪

行茶礼仪指的是在茶事活动过程中，茶者应遵循的礼仪规范。

1. 参与茶事活动者应专注泡茶，用心品茶。
2. 茶室、茶席、器具等应提前做好清洁准备工作。

3. 行茶动作应轻舒，避免器具之间发出较大的碰撞声。

4. 煮水时，壶嘴不朝向宾客或自己，应朝向无人的方向，防止沸水溅出烫伤。

5. 茶针、紫砂壶嘴、公道杯口等，尖的一面不应朝向宾客或自己，应指向无人的方向。

6. 回旋注水时，右手注水应按逆时针方向，有招手示意欢迎之意；若右手顺时针方向注水，手势似挥手则有不欢迎之意。

7. 出汤之前，应先持盖碗或壶在茶巾上吸干底部多余之水后，再移至公道杯出汤，防止壶底部的不洁净水滴入茶汤中。

行茶礼仪

8. 拿滤网、品茗杯时，注意手指不要触摸到品杯口或滤网内面，手指应拿住外部边缘1厘米以下位置或用茶夹夹取，以防止手指的细菌杂物混入到茶汤。

9. 分茶时，宜注七分满，不宜满杯。还需注意每位宾客分到茶汤量要均衡。

10. 奉茶的顺序遵循先长后幼，先主后从，女士优先等传统顺序，若无特殊主次之分，可从右至左顺序逐一奉茶。

二、普洱茶的品鉴要素

普洱茶愈久愈醇，愈久愈香，故普洱茶又被誉为"有生命的古董"。余秋雨先生曾在《普洱茶品鉴》一文中这样描述普洱茶："一团黑乎乎的粗枝大叶，横七竖八地压成了一个饼型，放到鼻子底下闻一闻，也没有明显的清香。扣下来一撮泡在开水里，有浅棕色漾出，喝一口，却有一种陈旧的味道。""香飘千里外，味酽一杯中"便是普洱茶的真实写照，小小的一片叶子，竟如此神奇。让我们一起融入这片小小的叶子之中，一步步走近普洱茶，学会认识、品鉴普洱茶。

近百年来，普洱茶深受广大消费者青睐，皆因茶质优良。同时普洱茶的独特风味，还与其自然陈化的过程有关，陈化后的普洱茶，经过特殊的加工程序，压制成大小不同、形状各异的茶团，置于干燥处自然阴干。再按运输要求，包装入篓，运住外地。历史上，云南地处祖国边疆，产茶区地处云南边陲，山高水险，在古代交通极为艰难，茶叶的外运全靠马帮牛帮，山路上耽搁的时间很长，有的路段马帮一年只能走两趟，牛帮则一年只能走一趟，茶在马背、牛背上长时间颠簸，使其内含物质徐徐转化，导致普洱茶的独特色泽更深、陈香风味更浓。普洱茶性较中和、正气，较适合香港人的肠胃，大多数人嗜饮，港九茶叶行商会理事长游育德先生把港人喜欢饮用普洱茶的原因概括为"五点"（十个字）：够浓，耐冲，性温，保健，价廉。

鉴别普洱茶，首先要明了其产地。普洱茶的原料为符合普洱茶产地环境条件的云南大叶种晒青茶，尤其以本书中详细介绍的名山所产之茶为优。有了好的原料，再经过好的加工，这样的结合便会诞生好的普洱茶。品鉴普洱茶的要素，可以从"色、香、味、形"四个方面来看。

（一）观色

品鉴普洱茶的第一步是观色，即一看茶色：不同种类的茶是以不同的形态呈现的，包括外形是否周正、芽叶嫩度、茶体颜色、压制松紧程度和茶体是否脱落等。不同茶的色泽、质地、匀齐度、紧结度、显毫状况也不相同；二看汤色：茶叶冲泡后，根据茶汤色泽的透亮度和清浊度也可分辨茶的品质；三看叶底：观看茶叶经过冲泡后，叶片的细嫩、匀齐以及完整程度，还要看其有无花杂和是否存有焦斑、红筋、红梗等现象。在购买普洱茶的时候不要简单地看包装、听宣传，更不要因为一些很会忽悠的卖茶人所说的而先入为主。

1. 看茶色

看茶色（即肉眼识茶），指的是看茶饼或干毛茶的色泽。好的普洱熟茶外形色泽褐红（称猪肝色），条索肥嫩、紧结。普洱茶从外形上分散茶和紧压茶。普洱散茶以嫩度划分级别，十级到一级、特级嫩度越来越高。

| 一级普洱熟茶 | 五级普洱熟茶 | 九级普洱熟茶 |

云南普洱茶主要是不同等级的茶叶拼配而成的，衡量外形主要看四点：一看芽头多少，芽头多、毫显，嫩度高；二看条索紧结、厚实程度，紧结、厚实的嫩度好；三是色泽光润程度，色泽就光滑、润泽的嫩度好；四是看净度，匀净、梗少无杂质者为好，反之则差。普洱茶紧压茶以散茶为原料，经蒸压成型的各种茶，花色品种众多，根据形状的不同有圆饼形的七子饼茶、有砖形的砖茶、有碗臼形的沱茶等等，各式各样大到几千克，小到几克，花色品种有上百种之多。鉴别普洱茶紧压茶的质量除内质特征与普洱茶散茶相同外，外形主要有如下要求：形状匀整端正；棱角整齐，不缺边少角；模纹清晰，撒面均匀，包心不外露；厚薄一致，松紧适度；普洱熟茶色泽以红褐、乌褐、棕褐、褐红色为正常；普洱生茶色泽以墨绿、褐绿、黄褐、棕褐等为主，随着贮藏时间的延长逐渐由绿转黄转褐。

鉴定普洱茶叶色泽是否正常，也是观色时所需要注意的。普洱茶叶色泽正常是指具备该茶类应有的色泽，如普洱生茶应黄绿、深绿、墨绿或青绿等；普洱熟茶外形色泽红褐、棕红、棕褐油润等。如果普洱生茶色泽显乌褐或暗褐，则品质肯定不正常；同样，普洱熟茶色泽如果泛暗绿色或呈现出花青色，品质也不正常。该条为评鉴普洱茶茶叶色泽的首要条件，其次看色泽的鲜沉、润枯、匀杂。生熟茶类评比色泽时注重色泽的新鲜度，即色泽光润有活力，同时看整盘茶是否匀齐一致，色泽调和，有没有其他颜色夹杂在一起。观茶色还包括看有没有霉梗、叶。晒青茶逢双取

普洱熟茶汤色

普洱生茶汤色

样，特级油润芽毫特别多，二级油润显毫，四级黑绿润泽，六级深绿，八级黄绿，十级黄褐。随着存放时间的推移，茶色会逐渐向褐绿、黄褐、棕褐方向转变。而普洱熟茶的茶色以褐红且均匀油润者为好，色泽黑暗或花杂有霉斑者较差。

2. 看汤色

看汤色（即开汤鉴茶），汤色也是制作工艺和茶的存放时间、存放条件状况的体现之一。普洱熟茶汤色要求红浓明亮。汤色晶莹剔透是高品质普洱熟茶的汤色，黄、橙色过浅或深暗发黑为不正常，汤色浑浊不清属品质劣变。当年的普洱生茶正常投茶量泡10余秒钟后汤色是黄绿色的，泡1分钟后汤色会变成金黄。存放一定年份的普洱生茶，其正常汤色变化应该是5年左右汤色绿色向黄绿转变，5~10年黄绿向金黄转变，10年后金黄向橙红转变。如果一款茶才有5~6年汤色就发生很大转变，不到10年就栗红就要怀疑是否经过轻度发酵或是进过湿仓。

鉴赏普洱茶的汤色时，最好选用晶莹剔透的无色玻璃杯（如鸡尾酒杯），向杯中斟入1/3杯的茶汤后，举杯齐眉，朝向光亮处，杯口向内倾斜45°，这样可以最精准地观察茶汤的色泽。鉴赏汤色的深浅、明暗时，应经常交换茶杯的位置，以免光线强弱不同而影响汤色明亮度的辨别。

3. 看叶底

看叶底，开汤后看冲泡后的叶底（茶渣），主要看柔软度、色泽、匀度。正常的普洱茶叶底是色泽一致，不软烂，无杂色，普洱生茶叶底随年份增加黄绿向金黄再向黄褐转变，而普洱熟茶的叶底随着年份的增加，颜色逐渐变黑褐。叶质柔软、肥嫩、有弹性的好，而叶底硬、无弹性的则品质不好；色泽褐红、均匀一致的好，而色泽花杂不匀，或发黑、碳化或腐烂如泥，叶张不开展的则属品质不好。

关于看普洱茶的叶底，可从以下两方面来品鉴：一是靠触觉辨别叶底的柔软度和弹性；二是靠眼睛判别叶底的老嫩、匀整度、色泽和开展与否，同时还观察有无其他杂物掺入。好的叶底应具备亮、嫩、软匀等因子。普洱生茶叶底呈黄绿色或黄色，叶条质地饱满柔软，充满鲜活感。也有些茶在制作工序中，譬如茶菁揉捻后，没有立即干燥，延误了很长时间，叶底也会呈深褐色，汤色也会比较浓而暗。普洱熟茶的叶底多半呈红褐色、暗栗色或黑褐色。有些熟茶若渥堆时间短，发酵程度轻，叶底色泽也会偏轻。此外，在赏鉴叶底嫩度时，要防止两种错觉：一是易把茶叶肥壮、节间长的某些品种特性误认为粗老条；二是湿仓茶茶色暗，叶底不开展，与同等嫩度的干仓茶比较，会被认为老茶。

晒青茶的叶底

陈年普洱生茶的叶底

普洱熟茶的叶底

（二）闻香

香气是茶叶的灵魂。香气是普洱茶永恒的魅力。有一首赞美普洱茶的诗曰："滇南佛国产奇茗，香孕禅意可洗心。"普洱茶的香气颇为丰富，普洱生茶香气是高雅幽远的，而普洱熟茶的香气是陈香显著且含蓄多变的。优质普洱茶的香型主要有蜜香、花香、清香、荷香、木香、樟香、兰香、枣香、陈香、药香等。辨别茶叶的香气，靠嗅觉来完成。通过冲泡茶叶，使其内含的芳香物质得到挥发，刺激鼻腔嗅觉神经进行区分，习惯上称为闻香。由于茶的品种、生态环境、树龄、纬度、海拔、土壤成分的差异，茶叶积累的物质也有差别，冲泡时挥发释放的香味香型、强弱也有区别。

樟香：云南各地有许多高大的樟树林，这些樟树多数高达数米，大樟树底下的空间最适合茶树的种植生长，大樟树可以提供茶树遮荫的条件，茶树在樟树环境下可以减少病虫害的发生。有人认为是茶树直接吸收了樟香贮存在叶片之中，于是普洱茶便有了独特的樟香。但实际上还是归因于茶树生长的光照、土壤等生态环境影响导致茶树生长代谢过程中产生了相应的香气前体物质，经特定加工工艺影响后最终使得普洱茶产生了独特的樟香。

荷香:"毛尖即雨前所采者,不作团,味淡香如荷,新色嫩绿可爱。芽茶较毛尖稍壮……女儿茶亦芽茶之类。""不作团"指的是不做成型的散茶;"味淡香如荷",描述了雨前毛尖非常幼嫩,茶汤很清淡,有莲荷香气;"芽茶较茶尖稍壮,女儿茶亦芽茶之类"荷香属于飘荡茶香,清雅娓娓。冲泡之前在赏茶时,可以从茶叶闻到淡淡荷香,冲泡工夫可直接影响普洱茶的荷香,宜用清新的好水冲泡,以软性水质最理想,冲泡时水温应沸热,以快冲速倒方式比较适宜。茶汤喝入口腔中,稍停留片刻,将喉头前的上颚空开,一股荷香经由上颚进入鼻腔中,在嗅觉感应下,散发淡然荷香,清雅娓娓,在叙说着普洱茶中浪漫情韵,激起了美之感性。

兰香:诗句"香于九畹芳兰气,圆如三秋皓月轮"是描述普洱茶兰香最美的诗句,"圆如三秋皓月轮",指的是像秋天圆大而美好的月亮般的普洱圆茶,"香于九畹芳兰气",1畹等于30亩,九畹是比喻广大而多,芳兰是指有香气的兰花,这一句的意义是形容比浓郁的兰花香更奇美。一般来说,条索较细长,色泽比较墨绿,叶底可明显看出是比较细嫩的茶菁,经长期醇化后,兰香较为浑厚,比较其粗老的茶菁加工而成的茶品醇化后的兰香更为清纯。

清香:在普洱茶中最为常见、典型的茶香。幼嫩的茶芽做成的茶品清香最为明显,香气不像花果香、蜜香等有独特的定位,而是展现香气清雅、原滋原味的感觉。清香型普洱茶主要看其原料和生产季节,一般来说,春茶和幼嫩茶清香更为明显,如"大益春早"清香非常突出。

枣香:枣香主要有青枣香与红枣香。青枣香属于花果香类别,一般在普洱生茶中某些区域茶品中含有,区分较为明显。红枣香则为普洱熟茶中含有,甜而带有红枣气味。一般而言,红枣香茶为适度发酵所致,新发酵出半成品较为清淡,但细品还是能闻出,产品经过后期陈化之后,红枣香将逐渐凸显出来。

蜜香:普洱茶之蜜香主要有三种:花蜜香、果蜜香、蜂蜜香。其中的花蜜香似花粉蜜,甜而刺激,花香中透露阵阵甜蜜。带有花蜜香的茶品一般经过一定的陈化后逐渐显露,如易武茶,后期陈化花蜜香更为显著。果蜜香甜而高雅,为普洱茶典型原香。勐宋那卡茶、景迈茶更为突出。蜂蜜香如蜂蜜散发出香味,一般在普洱熟茶中更为明显,主要由普洱熟茶发酵所致。有些陈化期较长的茶品中含有,特别陈化较长的普洱熟茶中更为明显。通过闻品茗杯更为显著,有"挂杯香"之说法。

花果香：普洱茶中常见的花果香有玫瑰花香、稻谷花香、兰花香、桂花香、梅子香等等，还有许多不知名而又特别突出野花香型。普洱茶之花果香类型非常广泛，且因不同区域环境及初制工艺，花香型亦不同。很多地域有典型的地域香，如班章老曼峨茶稻谷花香，布朗山、巴达山茶呈典型的梅子香，南糯山茶呈高雅的糯米香，景谷大白茶、格朗和茶呈玫瑰香，凤庆茶有典型的兰花型红茶原香等。

甜香：甜香主要有两种，一种为糖香，似焦糖或红糖香，透露出甜味；另外一种无糖的气息，纯粹为甜的气息。甜香主要为普洱熟茶的香型，在发酵过程中，因为大量的纤维素降解后形成的茶多糖、低聚糖及单糖所呈现的味道。而因发酵成熟偏重，有焦糖之香。在普洱熟茶品鉴过程中，甜香之意范围稍广，若能将甜香再进行细分，细细品来，产品甜香型便不同，亦可作为区分茶品的方法。

木香：茶品中的木香来自橙花叔醇等倍半萜烯类、4-乙烯基苯酚，主要由于茶品木质素的降解而产生的气味，似"木头味"，但韵味清幽、高雅而轻飘上扬。普洱茶中的木香主要来自于梗的陈化，一般带梗较多的茶品后期陈化较为突出，早期红印、绿印等产品木香特显。

陈香：品味陈香是普洱茶的至高境界与享受，每个喜爱普洱茶的茶友都知道普洱茶以陈为贵，越陈越香。陈香展现为时间的气息，历史的气息，似老酒醇活之底

蕴，轻淡而幽雅、低沉且缠绵，迷迭香绕，令茗者愉悦且沉醉。所以陈香是普洱茶随时间流逝而逐渐变化所散发出来的香气，闻时有让人迷醉的感觉。陈香在老茶中更为明显，气韵悠长，轻淡而缠绵，似有醉意。

"舒展皓齿有余味，更觉鹤心通杳冥。"繁华袭来，泡一杯普洱茶，把氤氲的意象泡开，闻着普洱茶香，可以让心远离尘嚣、心无旁骛。

在了解了普洱茶的十种主要类型的香型后，我们来谈谈品鉴普洱时的闻香技巧。

闻香气一般分为热闻、温闻和冷闻三个步骤，以仔细辨别香气的纯异、高低及持久程度。

1. 热闻

热闻是指冲泡出汤后立即趁热闻香气，此时最易辨别有无异气，如陈气、霉气或其他异气。随着温度下降异气部分散发，同时嗅觉对异气的敏感度也下降。因此热闻时应主要辨别香气是否纯正。

由于冲泡后的茶叶在热作用下，其内含的香气物质能充分地挥发出来，一些不良气味也能随热气挥发出来。所以，趁热湿闻叶底，最容易辨别出茶叶香气是否纯正，有无异杂味。其方法是一只手拿住已倒出茶汤的茶杯（壶或盖碗），另一只手半揭开杯盖（壶盖或碗盖），靠近杯（壶、碗）沿用鼻轻闻或深闻。为了正确判别茶叶的香气类型、香气高低、香气持续时间的长短，闻时应重复一二次，但每次闻的时间不宜过久，因为人的嗅觉容易疲劳，闻香过久，嗅觉的敏感性下降，闻香就不准确了，一般是2秒左右。未辨清茶叶香气之前，杯（壶、碗）盖不得打开。所以，当滤出茶汤或看完汤色后，应立即闻闻香气。

2. 温闻

温闻是指经过热闻及看完汤色后再来闻香气，此时评茶杯温度下降，手感略温热。温闻时香气不烫不凉，最易辨别香气的类型、浓度、高低，应细细地闻，注意体会香气的浓淡高低。

3. 冷闻

冷闻即闻杯底香，是指经过温闻及尝完滋味后再来闻香气，此时评茶杯温度已降至室温，手感已凉，闻时应深深地闻，仔细辨别是否仍有余香。如果此时仍有余香是品质好的表现，即香气的持久程度好。

热闻、温闻、冷闻三个阶段相互结合才能准确鉴定出茶叶的香气特点。每个阶段辨别的重点不同（表4-3）。

表4-3 普洱茶香气辨别方法和技巧

辨别方法	辨别的重点	注意事项
热闻	香气纯异度、香气类型和高低	叶温65℃以上时，最易辨别茶叶是否有异味
温闻	主要辨别香气类型和茶香的优劣	在叶底温度55℃左右，最易辨别香气类型
冷闻	主要辨别茶叶香气的持久程度	叶温30℃以下时，辨别茶香余韵，高者为优

鉴赏普洱茶的香气是怡情悦志的一种精神享受。为了更好地闻香，宜选用较大的柱形瓷杯做公道杯。因为瓷质器皿的内壁比玻璃器皿更容易挂香，而且杯的内积大，可聚集更多的茶香，让茶香更饱满、更丰富，可以更准确地鉴别茶香的优劣。

最后，饮尽杯中茶，再闻一闻杯底留香，借以判断香气的持久性和冷香的特征。好的普洱茶香气纯正细腻，优雅协调，可令人心旷神怡，杯底留香明显而持久。所谓的杯底香，往往是茶人们的最爱，他们喝茶后，不把杯子放下，而是在手中把玩，细细品味那残留在杯子底下的最后那丝丝香气。而杯底香中以冷杯香最玄，冷杯香是汤冷了以后的香气。有人会问，茶冷了以后还有香气吗？不知道在你喝茶的时候注意了没，杯底香往往跟汤底香是紧密联系在一起的，这是表面香之外的香，藏在汤里，驻在水中，挂在杯上。有温度的时候，香会明显一些，没有温度的时候，香会收敛一些，但始终在发散，绵绵密密，若有似无。有时会觉得忽然强势张扬，但仔细去寻找，又无影无踪。

在普洱茶气味鉴别中，主要区别霉味与陈香味。有人说陈香味是霉味，这是错误的。霉味是一种变质的味道，使人不愉快，不能接受的一种气味。而陈香味是普洱茶在后发酵过程中，多种化学成分在微生物和酶的作用下，形成了一些新的物质，这些新的物质所产生的一种综合的香气，犹如老房子的感觉，是一种令人感到舒服的气味。如乌龙茶中的铁观音有"余韵"，武夷岩茶有"岩韵"一样，普洱茶所具有的是"陈韵"。"陈韵"是一种经过陈化后，所产生出来的韵味，只能意会不能言传，但能引起共鸣、领会，激起思古之幽情，引发历史之震撼。这是普洱茶香气的最高境界，普洱茶纯正的香气是具有陈香味和以上所说的几类香气。有霉味、酸味、馊味和刺鼻的味道都为不正常，要谨慎饮之。

（三）品味

古语云"味之有余谓韵"。韵味是茶汤中各种呈味物质比例均衡，入口爽快舒适,滋味厚重馥郁又具有层次变化,让人愉悦地感受到某种超越味道的感觉。这是普洱茶能带给人们的更深层次的享受。这种感觉或许能让人们在品茗过程中感受到某种美好的意境,而此意境既使心灵净化,又使人自在超脱。

陈年普洱茶在陈化过程中的糖化作用,使得茶体转化出的单糖又氧化聚合成多糖,使得其汤入口回甜,久久不去,喉头因此润化,渴感自解。饮用陈年普洱茶能达到舌面生津的效果,茶汤经口腔吞咽后,口内唾液徐徐分泌,会感觉舌头上面非常湿润,这种感受比较独特。相反,质量不佳的普洱茶,茶汤入口会觉得喉头难受,产生干而燥的感觉,强烈者甚至影响吞咽。

口感,是味觉、嗅觉、触觉对茶叶茶汤产生的各种刺激所形成的综合的主观感受。普洱茶的口感源于其水浸出物,而茶叶的本质是基础。通常普洱茶水浸出物为30%～50%,不同类别的物质的口感各有其特性。

1. 茶多酚类

茶多酚类物质的口感主要表现为涩味,有收敛性。茶多酚的化学组成复杂,以儿茶素类物质为主,可分为酯型儿茶素和非酯型儿茶素,前者刺激性较强,涩味明显,并使口腔感觉"粗糙";后者刺激性弱,使口腔感觉"爽口""顺滑"。

品质好的茶入口"抓"舌头,但很快松开,这种感觉被称为"化",这样的茶即便在茶汤温度降低后也不会留有过重的涩底。有茶人把"抓"舌头的力度、"化"的时间长短作为评判茶叶品质的依据之一。

2. 生物碱类

表现出的口感是苦。苦味在口中的刺激程度以及散化的快与慢也是判断茶叶品质的因素之一。苦味不散或是过于强烈都会让人反感。

3. 氨基酸类

表现多样，他们与其他呈味物质有很强的协同作用。鲜味：茶氨酸、谷氨酸、天冬氨酸；甜味：甘氨酸、丙氨酸等；酸味：谷氨酸、天冬氨酸等；香味（花香）：谷氨酸、丙氨酸。

4. 糖类

在味觉表现为甜味，在嗅觉为甜香味。甜味对口感有很大影响，在人的本能需要中糖是最首要的，味觉及嗅觉对甜味都非常敏感，甜味能让人产生愉快的感觉。糖类中的果胶对口感有重要的作用。果胶在嫩度适中的茶叶中含量最高达干茶的3%～5%，有黏性，能让口腔感觉"稠""滑"。陈化过程中果胶可降解为水溶性碳水化合物从而增加滋味。

5. 芳香物类

呈现香味。不同品种、不同产地的茶叶所呈现的香味有很大差异，随着存储年份的增加香味由清香→花香→果香→蜜香→木香→陈香转化，香气的挥发性由"扬"变"稳"，芳香物的化学成分也出现分子结构逐渐加大的规律，这是陈化过程中化合物氧化聚合所致。古树茶通常呈现出花、蜜香，而且杯底也留有明显的蜜、甜气味，而且这类茶叶在存储时转化较快。

6. 其他口感

生津——直接的原因是茶汤中各种化合物刺激口腔而兴奋唾液分泌中枢所致。往往在停止刺激后唾液还会持续分泌，这更多是由于茶氨酸兴奋副交感神经的缘故。副交感神经兴奋会导致较为持续的胃肠蠕动及唾液分泌。唾液的持续分泌让口腔及咽喉润泽，唾液腺中的黏液性腺泡所分泌的糖蛋白与茶汤中其他呈味物质的协同作用而产生"回甘""润喉"的感觉。唾液不但可以保持口腔清洁，还有帮助消化、保护胃黏膜的作用。古人非常重视唾液与养生的关系，美其名曰"琼浆玉液"。

普洱茶通常有甜、苦、涩、酸、鲜等数种味感，也有滑、爽、厚、薄、利等口感，同时还有回甘、喉润、生津等回感。此类味感、口感、回感等组合而成普洱茶之滋味，各种感觉可能单独存在某一泡普洱茶中，也可能并存，在滋味品鉴过程中就需要细细品味。

（1）味感

甜：甜的感觉绢细而绵长，让人感觉甜而不腻。甜味是由碳水化合物经水解或裂解形成可溶性单糖或低聚糖相关。

甜味不仅是儿童喜欢，成年人也都会对糖而垂涎。但是浓糖甜腻，往往使人又爱又怕，然而茶中的淡然甜意是那么清雅，对健康无害。由于淡然甜意，更将普洱茶品茗提升到艺术境界。普洱茶原料为云南大叶种，内含物质丰富，经过长期陈化，苦和涩的味道因氧化而慢慢减弱，甚至完全没有了，而糖分仍然留在茶叶中，经冲泡后，慢慢释放于普洱茶汤里，而有甜的味道。普洱茶汤中的甜味，纯正清雅，也最能代表普洱茶的真性。

苦：苦本是茶的原味，古代称茶为苦茶，早已得到了印证。先苦而后才能回甘，并带给普洱茶品茗者那种真道的启示。普洱茶之所以会有苦，是因为其中含咖啡碱，茶所以能提神醒目，就是因为这些咖啡碱，对人体神经系统引起了兴奋作用的效果。真正健康的普洱茶品茗，并非透过苦味去求得提神醒目，而是从略带苦意的茶汤，达到回甘喉韵之效。以比较幼嫩等级的茶菁所制成的普洱茶，都带有苦味。至于对苦味的处理，都是以冲泡方法来控制，同时也视各品茗者对苦味的接受程度，而泡出适当的苦味茶汤。

好的普洱茶有很明显的"先苦后甜"感，回甘是普洱茶的一大特征，也是人们喜欢饮普洱茶的一个原因，回甘强弱与持久度是鉴别一款茶的因素之一。像著名的老班章、景迈这些名山古茶，饮茶后如果没吃其他东西干扰味觉，口腔咽喉的甜滑感可以持续一两个小时。

涩：由酯型儿茶素与口腔细胞中蛋白质发生络合造成，感觉舌苔增厚，口腔内壁增粗，有东西黏附。常听说不苦不涩不是茶，其实陈化后的陈老普洱茶，苦涩逐渐褪去。普洱茶有口感比较强的阳刚性普洱，有口感比较温顺的阴柔性普洱。哪些是刚性的？哪些是柔性的？以其苦涩的程度而定，是最具体辨别方法。茶的涩感是因为含有茶单宁成分，普洱茶是大叶种茶菁制成的，所含的茶单宁成分比一般茶叶多，所以新的晒青茶滋味十分浓强，也使涩的口感特强。适当的涩感对品茗者是可以接受的，因为涩会使口腔内肌肉收敛，而产生生津作用。涩可以增加普洱茶汤的刚强度，也满足口感较重的品茗者，冲泡苦味和涩味都需注意其技巧与个人接受度。

鲜：鲜包含鲜爽、鲜活、鲜甜、鲜美、鲜灵。是由于氨基酸类及茶黄素，还有微量的可溶性的肽类、核苷酸、琥珀酸等成分造成。

酸：酸味是普洱茶不好的味道，当然在普洱茶品茗时不希望有酸味出现。酸味是不好的品质所产生味感，会令牙齿及两颊紧张、收拢，是由原料干燥不及时，发酵过程中堆温不当或保存不当等造成。

（2）口感

口感是在基于味觉的基础上综合口腔内的其他感受神经共同体会并作出综合感受评价。口感的形成不仅受茶汤的化学组成影响，还同时受到茶汤密度、黏稠度、温度等一系列的物理因素的影响。

滑：从感官上来形容，就像流体轻轻的拂过你的舌面，进入喉咙，是一种舒顺与安逸的感受。影响普洱茶顺滑物质主要的普洱茶中的可溶性糖类，寡糖与果胶。糖类物质能包裹住多酚类物质，减弱"涩"感，增加茶汤"滑"感。普洱茶的品鉴中，滑是对普洱茶品质鉴别的重要指标，不顺滑的茶汤要么"卡喉"，要么清淡，给人以不舒适感。一般老茶或熟茶，顺滑较明显，而晒青茶特别是新品滑感很弱。

化：指茶汤在口中流转时滋味感觉变化的速度。化的描述比较抽象，一般品茗者在品饮时所说的化与不化是茶汤入口后在口腔味觉上停留的时间。"入口即化"是指茶汤的滋味在进入口腔停留数秒后能够自然消散，而回味无穷；"入口难化"的茶汤则是其滋味久久停留在舌苔上难以散去。

活：活主要指在口腔中产生一种活泼、愉快、力量的感受，与"滑"相比，活的口感给品鉴者的体会更加清新与灵动。历经时光锤炼的普洱茶，内含成分经过复杂的转变，冲泡成茶汤后，才有较强的活性品味。经过醇化的干仓普洱茶，内含成分经过水解、裂解、氧化等各种化学反应，水溶性物质增加，分子量降低，物质在茶汤中转化迅速，也能有最好的活性品味表现。

水：原料粗老，呈味物质较少或工艺不当中茶体所处位置湿度太高导致内含物流失，均会造成茶汤"水"的感受。其他味道化去之后，有回寡之感，就像喝杯清水，寡淡无味。

厚与薄：茶汤厚说明茶汤内含物丰富，水浸出物含量高，耐泡，茶汤厚重，有质感，俗语"能够压住舌头"。薄则是茶汤淡薄轻飘，而且比较寡淡，内含物质较单一，水浸出物含量低，整个茶汤呈现薄、不协调。

利：俗称刮喉。主要是因为茶汤中内含物质不协调。茶汤中有些内含物质太多，有些内含物质太少，不能完美地平衡各种复杂的味道，使得一种或几种偏激、浓烈的味道过度地刺激味觉与触觉，使品鉴者感受像利刃在喉，收刮我们的喉咙。

（3）回感

普洱茶的味感与口感是品茗者的真实感受与体会，但普洱茶的品饮不仅仅只有这些感受与体会，还会出现饮后回味。

主要包括：回甘、喉润与生津三部分，此等反应是茶品给予品鉴者的礼物，也是饮后感觉的升华，心灵的享受。

回甘：甘似甜，但不同于甜，甜为茶汤浸润舌尖而有甜味，而回甘则是品茗者在品味茶汤过后，自身形成的甘甜感受。回甘的体会比较内敛，细腻而绵长。品茗者在品饮茶汤后，口腔内出现丝丝甜意就是回甘的表现。

回苦：回苦与回甘相反，饮后，苦味依旧，转至喉口，久化不去。苦味在普洱茶品鉴中有两种，一种为入口即苦，苦化为甘，也称先苦后甜，另一种为茶汤入口不苦，后化苦，久久不散。

润：润的体现如同回甘，是在品饮茶汤后出现润滑、滋润感觉，可以说润是在品茗者对于茶汤品味后的综合反映。润的体会不但说明茶汤滋味饱满，而且口感湿滑不出现卡、刮喉现象，品茗者适应茶汤的滋味、口感后才会形成润的体会。

生津："津，唾液也"。生津也就是口腔中分泌出唾液之后的感觉。普洱茶的原料为云南大叶种晒青毛茶，茶叶内含成分丰富，特别是酯型儿茶素（EGCG、ECG等）含量高，由涩而生津，生津功能特强。部分较劣等茶品，品饮后始终觉得口腔内部卷起，两颊肌肉痉挛般难受，舌苔增厚，但无生津之感。这种涩而不能生津，称之为"涩化不开"。生津具体细分为两颊生津、齿颊生津、舌面生津、舌底鸣泉（舌下生津）等。

①两颊生津

两颊生津为生津中最为激烈的一项。茶汤入口后，因为呈涩物质刺激口腔两侧内膜而分泌出唾液，因此造成生津是属于"两颊生津"。

两颊生津所分泌的唾液，通常是比较多而强。这种生津在口感上，会觉得比较粗野且急促，口中有大量唾液，挤满整个口腔，从而使生津之感非常强烈。早春茶或幼嫩涩感较足的茶品两颊生津较为明显，体内失水过多，多选具有两颊生津效果的茶品，冲泡饮用解渴效果特好。

②齿颊生津

品饮普洱茶过程中，茶汤在口中流动，单宁类物质刺激两颊与牙齿之间内膜，促使分泌唾液而产生生津。齿颊生津与两颊生津虽然生津位置不同，感觉更是不同。两颊生津如瀑布洪泄，粗野而急促；齿颊生津则如涓涓溪流，柔细而绵长，浸润之处，温润而甘滑。齿颊生津在普洱熟茶轻发酵工艺产品品饮过程中感觉较为鲜明，饮后，齿颊之间如绵长溪水，丝丝甘泉，余水不绝。

③舌面生津

在品饮过程中，涩感化得较快的茶品，饮后在舌面上会有层湿润的浆液，从而产生舌面生津的现象。茶汤经口腔吞咽后，口内唾液徐徐分泌出来，在舌头的上面，非常的温润柔滑、缓和细致，同时，舌面好像在不断地分泌出唾液，然后流到舌头两边口腔。好的普洱茶，基本都能达到舌面生津的效果。

④舌底鸣泉

茶汤进入舌底与下牙床交替处，因生津而感觉有"泡泡"冒出，这样的现象，也称"舌下生津"。品饮醇化时期较长的普洱老茶，茶汤经过口腔接触到舌头底

部，舌头底面会缓缓生津，会不断有涌出细小泡泡的感觉。这是因为茶多酚在醇化过程中，经氧化、水解、合成、裂解等大规模的化学反应，已经不能刺激两颊或舌面生津，但是新合成的一些物质成分，起到激起舌底鸣泉的作用。舌底鸣泉生津过程更加缓和持续，生津现象更加细致轻滑，生津感受更加柔顺安详。在品饮陈年普洱茶的时候，茶汤极为柔和，经过口腔接触到舌头底部，舌底会缓缓生津，仿佛不断涌出细小的泡沫，这种舌下生津的现象，才是真正的舌底鸣泉。

想要品出普洱茶汤之味，是需要讲究些技巧的。切忌像喝饮料一样的"牛饮"，这样连茶是什么滋味都还未尝到，就已经喝饱了。大体的原则是：小口慢饮，口内回转，缓缓咽下。茶汤入口之时，应将口腔上下尽量空开，闭着双唇，牙齿上下分离，增大口中空间，同时口腔内部肌肉放松，使舌头和上颌触部的部位形成更大的空隙，茶汤得以浸到下牙床和舌头底面。吞咽时，口腔范围缩小，将茶汤压迫入喉，咽下。在口腔缩小的过程中，舌头底下的茶汤和空气被压迫出来，舌底会有冒泡的感觉，这种现象就叫做"鸣泉"。品茶要品出境界，贵在茶好水好之外，还要有一种品茶的好心情，才能凝精聚神地穿透茶的本质，提升到感悟的精神意境。

在掌握了这些简单的品尝方法后，再次品尝普洱茶，便可品出普洱茶陈香所透露出的深厚历史韵味，彰显着返朴归真的自然真性。

（四）耐泡度

耐泡度指的是茶经过多次冲泡后，其汤色口感没有太大的变化。而由于茶类的不同，其耐泡程度不一样。人们日常生活中，常有这样的体会，人们饮用的袋泡红茶、绿茶及花茶，一般应该冲泡1次后就将茶渣弃掉了。因为这种茶叶在加工制造时通过切揉，充分破坏了叶细胞，形成颗粒状或形状细小的片状，茶叶中的有效成分冲泡时很容易被浸出来。普通绿茶常可冲泡3~4次。铁观音之类的乌龙茶素有"七泡留余香"之美名，即只能泡七八泡。而普洱茶，应该是诸多茶品里最耐泡的茶了，它之所以耐泡，是普洱茶所含丰富的内含物质。普洱茶历经了数百上千年的生长，它的芽叶上积攒了丰富的营养物质，在饮用时必然要经过很多次的冲泡才能释放完毕，这就是我们感觉到的它经久耐泡的缘故。在正常冲泡下，普通普洱茶能冲泡十至二十泡，而且在每一次的出汤都会有一些奇妙的变化。

总而言之，普洱茶滋味有甜、酸、苦、涩、鲜之味感，醇、厚、滑、薄、利之口感，回甘、喉润、生津之回感。生津之趣亦妙，两颊生津如瀑布洪泄，粗野而急促，齿颊生津如涓涓溪流，柔细而绵长；舌面生津如温润甘露，娇柔而细致，舌底鸣泉如丝丝清泉，轻滑而安祥。品鉴过程亦享受过程，质、量、度、时间的把握，就能品出真味。

三、普洱茶的审评

在之前介绍普洱茶的品鉴要素时已提出，普洱茶的种类繁多，依制法分为普洱生茶和普洱熟茶；依存放方式分为干仓普洱茶和湿仓普洱茶；依外型分为饼茶（七子饼）、沱茶、砖茶、散茶等。接下来，我们将从看干茶、观汤色、闻茶香、品滋味方面来介绍普洱茶的审评。

（一）看干茶

主要从干茶形状、整碎、色泽和净度来评审。

形状：包含茶品的外形规格，如大小、长短、粗细、轻重、压制的形状、松紧度、匀整度等。

整碎：指条索的大小、长短和粗细是否均匀、完整；上中下各段茶比例是否匀称。

色泽：指茶品的颜色的深浅程度和光泽度，茶品色面的亮暗和油润程度（深浅、润枯、鲜暗、匀杂等）。

净度：指茶类夹杂物（梗、籽、朴、片等）、非茶类夹杂物（杂草、树叶及其他）的含量。

总之，从外形来看，好的普洱茶外形的条索结实、颜色鲜润、油面光泽，充分表现了茶叶的活力感。

外形常用术语：

端正：指形态完整，无破损残缺，整齐。

松紧适度：指压制茶松紧适当。

平滑：指表面平整，无翘起、脱皮及茶梗刺出等现象，反之成称为粗糙。

锋苗：芽叶细嫩，紧卷而有尖锋。

重实：身骨重，茶在手中有沉重感。

壮结：茶条肥壮结实。

轻飘：身骨轻，茶在手中分量很轻。

粗松：嫩度茶，形状粗大而松散。

芽头：指未发育成茎、叶的嫩尖，质地柔软，茸毛多。

茎：未木质化嫩梗。

梗：着生芽叶的已木质化嫩枝，一般指当年青梗。

金毫：嫩芽带金黄色茸毫。

显毫：茸毛含量较多。

猪肝色：红而带暗，颜色似猪肝。

棕褐：褐中带棕。

褐红：红中带褐。

黑褐：褐中带黑。

褐黑：乌中带褐，有光泽。

黑润：色黑而深，似涂上一层油而亮。

乌黑：深黑色。

（二）观汤色

茶品汤色审评主要从色度、亮度和清浊度三方面去评比。汤色审评要及时，因为溶于热水中的多酚类物质与空气接触后很容易氧化变色。

色度：即茶汤的颜色类型和深浅。与茶树品种和鲜叶老嫩有关；加工工艺决定了各类茶不同的汤色。

亮度：指茶汤明暗的程度。凡茶汤亮度好的品质亦好。

清浊度：指茶汤的透明程度。汤色透明无杂质，清晰透亮；汤色浑浊，漂浮杂质较多，浑不见底。

接下来我们就来具体看看普洱熟茶、普洱生茶的汤色的辨析方法。

首先来看看普洱熟茶的汤色分类。

橙红：红中带橙。

深红：红而深，缺乏明鲜光彩。

栗红：红中带深棕色，也适用于普洱熟茶的叶底色泽。

红浓：汤色红而深浓，茶汤颜色红，且内含物丰富。

褐红：红中带褐。

红褐：褐中带红。

而普洱生茶和晒青茶的汤色分类又有所区别，主要有：

黄绿：以绿为主，绿中带黄。

绿黄：以黄为主，黄中带绿。

嫩黄：金黄中泛出嫩白色。

浅黄：内含物不丰富，黄而浅。

深黄：黄色较深，无光泽。

黄亮：色黄、有光泽。

橙黄：黄中微带红。

总之，好的普洱熟茶汤色是红浓明亮的，好的普洱生茶和晒青茶是黄亮的。

（三）闻茶香

审评香气除辨别香型外，主要评比香气的纯异、高低和长短。

纯异：指香气的纯正度。异指茶香不纯或沾染了外来气味，如烟焦味、酸馊味、油味等。

高低：主要从浓、鲜、清、纯、平、粗进行评审。

长短：香气的持久程度。长指从热闻到冷闻都能闻到香气；反之则短。

香气常用术语：

毫香：芽毫显露的茶品所具有的香气。

清香：香清爽鲜锐。

幽香：香气幽雅，似花香。

花果香：似新鲜花、成熟果香气。

焦糖香：烘干充足或火功高致使香气带有糖香。

甜纯：香气纯而不高，但有甜感。

馥郁：香气幽雅，芬芳持久。

浓烈：香气丰满持久，刺激性强烈。

（四）品滋味

良好的味感是构成茶品品质的主要因素之一。茶滋味与香气关系密切。评茶时能嗅到的各种香气，如花香、熟板栗香等，往往在评茶滋味时也能感受到。一般说香气好，茶滋味也是好的。茶香气、茶滋味鉴别有困难时可以相互辅证。审评茶滋味适宜温度在50℃左右，主要区别其浓淡、强弱、鲜、爽、醇、和等。

浓淡：浓指浸出的内含物丰富，有黏厚的感觉；淡则相反，内含物少，淡薄无味。

强弱：强指茶汤吮入口感到刺激性或收敛性强，吐出茶汤后短时间内味感增强；弱则相反，入口刺激性弱，吐出茶汤后口味平淡。

鲜爽：鲜似食新鲜水果感觉，爽指爽口。

醇：醇表示茶味尚浓，回味也爽，但刺激性欠强。

和：表示茶滋味平淡正常。

滋味审评术语：

浓厚：入口浓，刺激性强而持续，回甘。

醇厚：入口爽适甘厚，余味长。

醇和：醇而平和，回味略甜。刺激性比醇正弱而比平和强。

平和：茶味正常、刺激性弱。

平淡：入口稍有茶味，无回味。

水味：茶汤浓度感不足，淡薄如水。

回甘：茶汤饮后在舌根和喉部有甜感，并有滋润的感觉。

鲜爽：新鲜爽口。

青涩：茶味淡而青草味重。

苦底：入口即有苦味，后味更苦。

（五）看叶底

审评完滋味后，将叶底倒入叶底盘中，观察其嫩度、匀度、色泽。叶底的老嫩、匀杂、整碎、色泽的亮暗和叶片展开的程度等是评定茶品优次的重要因素。好的叶底应具备亮、嫩、厚、软等几个或全部因子。

叶底常用的品鉴术语：

褐红：红中带褐，也适用于普洱茶渥堆正常的干茶色泽。

红褐：褐中带红，为普洱茶渥堆成熟的叶底色泽。

绿黄：以黄为主，黄中泛绿，比黄绿差，也适用于汤色。

黄绿：以绿为主，绿中带黄。此术语也适用于汤色。

花杂：叶色不一，形状不一或多梗、朴等茶类夹杂物。

嫩软：芽叶嫩而柔软。

嫩匀：茶品嫩而柔软，匀齐一致。

相信经过以上的品鉴和审评介绍，再加上大多数普洱茶品茗者的亲身体会，就更能体味普洱茶每一泡的滋味，真正感受普洱茶的魅力。

第四节 云南名茶山和名茶品鉴

一、云南名茶山介绍

云南的名茶山主要分布在云南五大产茶区，即西双版纳茶区、普洱茶区、临沧茶区、保山茶区和文山茶区。每个茶区都有它不同的特色，各座茶山树木林立，鲜花盛开，终年小溪叮咚流淌，蝴蝶起舞飞翔。也因每座茶山的地理位置不同，水质气候相异，每个山头上茶的滋味也不一样。现将主要茶山介绍如下：

（一）老班章

老班章位于云南省西双版纳州勐海县布朗山乡，海拔在1600米以上，最高海拔达到1900米，平均海拔1700米，属于亚热带高原季风气候带，冬无严寒，夏无酷暑，一年只有旱湿雨季之分，雨量充沛，土地肥沃，有利于茶树的生长和养分积累。老班章在

老班章

布朗山乡政府北面，为哈尼族村寨，有古茶园4490亩。自古以来，老班章村民沿用传统古法人工养护赖以为生的茶树，遵循民风手工采摘鲜叶，土法炒制揉作茶箐。老班章普洱茶，茶气刚烈，厚重醇香，霸气十足，在普洱茶中历来被尊为"王者""茶王"等。老班章普洱茶干茶条索肥壮，滋味浓厚，质重气强，苦味重但化得很快，且回甘迅猛持久。香型在兰花香与花蜜香之间，且非常饱满，杯底留香，馥郁持久，非常迷人。此外，老班章十分耐泡，二十多泡之后，茶味依然香浓。

（二）冰岛

冰岛村隶属双江拉祜族佤族布朗族傣族自治县勐库镇，地处勐库镇北边，距勐库镇政府所在地25千米，距县城44千米。东邻临沧，南邻坝卡，西邻耿马，北邻临沧。辖冰岛老寨糯伍、坝歪、地界、南迫5个村民小组。现有农户273户，有乡村人口1064人，其中农业人口1064人，劳动力955人，海拔1400～2500米，年平均气温18～20℃，年降水量1800毫米，主要民族为傣族、拉祜族、布朗族。

冰岛是勐库大叶种的发源地，种植历史悠久，最早有种茶的历史可追溯到明成化年间。勐库大叶种是国家级良种茶，被称为大叶种的正宗、大叶种英豪。

冰岛五寨的古树茶风格各异，东半山两个寨子坝歪、糯伍高香、苦味轻，涩感较明显，但生津较好，回甘持久，汤略薄。西半山3个寨子老寨风格明显，古树无明显苦涩，香气高扬，茶汤饱满，生津快，回甘快且持久，并以具有独特的冰糖韵而出名。地界及南迫则香气好，苦味较其他寨子稍重，无明显涩感，生津稍差，但回甘较好，滋味饱满。

冰岛

（三）昔归

昔归村隶属于云南省临沧市临翔区邦东乡邦东行政村，北纬23°~25°，海拔704~1043.4米，属于典型的低纬度山区。距离村委会12千米，距离乡政府16千米。面积3.82平方千米，年平均气温21℃，年降水量1200毫米。

昔归古茶园多分布在半山一带，混生于森林中，属邦东大叶种，是勐库冰岛茶

昔归

的一个分支，古茶树中较大的茶树基围在60~110厘米。清末民初《缅宁县志》记载："种茶人户全县约六七千户，邦东乡则蛮鹿、锡规尤特著，蛮鹿茶色味之佳，超过其他产茶区。"这里说的蛮鹿，现称为忙麓，锡规现称为昔归。忙麓山的茶还有一个特点，是自然生长的。有的树高三四米，有的五六米，有几棵茶树主干只剩下一截枯树桩，但又从底部重新长出了锄把粗的新树叉。大茶树基围在80~90厘米左右，茶园属传统采摘自然生长，树枝盘曲向上，经百年的人工无意造作，形成的造型嶙峋古怪，似卧龙、似飞禽展翅，既易攀援采摘又有观赏性，是典型的人工

栽培古茶园。

昔归茶内质丰富十分耐泡，茶汤浓度高，滋味厚重，香气高锐，茶气强烈却又汤感柔顺，水路细腻并伴随着浓强的回甘与生津，且口腔留香持久。

昔归茶开汤，汤色淡黄清亮，入口即香，无杂味，味甘；三泡后回甘更明显，香气高锐，两颊与舌底生津，舌面感觉微涩，化得很快；四至六泡，香气如兰，冰糖香渐显，水质较黏稠，重手泡后苦现，较轻，易化；七泡后汤色几乎未变，醇厚，更佳，尚微涩，喉韵深，回味悠长；十泡后水渐淡，甜味稍减，回甘好，冰糖香尚存。叶底墨绿柳条形，柔韧光鲜。

（四）易武

易武属于云南省西双版纳州勐腊县。易武茶山位于六大茶山的东部，紧靠中老边境，面积约750平方千米，是古六大茶山中面积最大、产量最高的茶山。易武乡拥有古茶园面积1.4万余亩，主要集中在高山寨、落水洞、麻黑、曼秀、三合社等村寨。易武乡北与普洱江城接壤，南接瑶区、勐伴，西接勐仑象明，东邻老挝。海拔差异大，气候立体型，不同小区气候条件，造成了不同的生态环境，使之具有温暖、较温暖型两种气候特点。易武常年日照充足，雨量充沛，全区山高雾重，土地肥沃，温热多雨，热量丰富，雨量充沛，是种植茶叶的理想之地。易武山高雾重，土地肥沃，温热多雨，热量丰富，雨量充沛。茶区土壤，在热带亚热带季雨林成土条件下，由紫色岩和沙岩母岩上风化发育而成，主要为砖红壤、赤红壤、黄壤。各地土质呈微酸性反应，pH值在4.5~6.5之间。土壤养分积累快，分解利用快，土壤有机质含量4.5%以上，腐殖质厚5厘米以上。土层深厚，土壤透气性好，有机质含量高。古茶树分布区域植被生态系统保持较好，生长着诸如椿树、香樟树、榕树、漆树、董棕等高大乔木。气生植物多，树木、藤本植物园繁茂，森林覆盖率高和高等植物集中，构成了良好的生态环境，是种植茶叶的理想之地。

易武茶素以香扬水柔、柔中带刚著称，干茶条索肥壮紧结，叶长色润，花蜜香

易武

馥郁持久，汤色橙黄明亮，且温润柔雅，茶汤入口甜滑、协调性好、口感变化层次丰富，苦涩度低，刺激性弱，汤感细腻，回甘生津，韵味十足，且耐泡度较高。

（五）勐库大雪山

在临沧市双江县西部与耿马县交界处，有一座南北走向的横断山系支脉——邦马山。主峰叫勐库大雪山（临沧地区还有永德大雪山、邦东大雪山），海拔3200多米，位于双江县勐库镇境内。著名的勐库野生古茶树群落就位于此山海拔高度2200～2750米的地方。勐库野生古茶树群落是目前国内外已发现的海拔最高、密度最大的野生古茶树群落，分布面积约12000多亩。勐库大叶种茶树叶片特大，长椭圆或椭圆型，叶色深绿，叶质厚软。芽叶肥壮，黄绿色，绒毛特多。春茶一芽二叶干样约含氨基酸1.7%，茶多酚33.8%，儿茶素总量18.2%，咖啡碱4.1%。适制红茶、绿茶和普洱茶。勐库大雪山普洱茶口感厚重，汤质稠厚，香气深沉，内质丰富。

勐库

（六）老曼峨

老曼峨自然村隶属于云南省勐海县布朗山乡班章村委会行政村。位于布朗山乡东北边，距离布朗山乡政府16千米。土地面积68.4平方千米，海拔1650米，年平均气温18～21℃，年降水量1374毫米。它是整个勐海县布朗山最古老、最大的布朗族村寨。据寨里古寺内的石碑记载，其建寨时间恰好就是傣族传统的傣历元年纪年，至今已有1370年的悠久历史。这里的古茶园中，一棵棵刻满沧桑岁月的古茶树，见证了布朗族先民"濮人"久远的种茶历史。

老曼峨

老曼峨茶树分布村落四周，是勐海大叶种的典型代表，主要以栽培型古茶树为主，茶树树龄在100～500年。老曼峨也有才栽种几十年的小茶树。

苦是老曼峨茶的一大特征。春茶性极寒、品饮苦若黄连，条形肥壮厚实、匀称显毫，汤色黄明透亮，有明显苦寒气息，滋味浓烈厚实、久泡有余香、耐冲泡，入口苦味比较重，略带涩感，但苦涩化得快，且持久。相对苦茶而言，老曼峨甜茶香气更醇，更高，但还是有明显的苦涩感，化得较快，回甘很好，生津也比较好。

（七）坪寨

坪寨古茶山位于法斗乡境内，地处云南省文山壮族苗族自治州西畴县东南部。北纬23°16′~23°32′，东经103°23′~104°92′，其核心范围完全坐落于北回归线之上；茶山面积约109.7平方千米。其地处华南丘陵与云贵高原的交接地带，常年云雾缭绕，茶区年雾天数达146~198天，低纬度高海拔的条件为茶树提供了充足的光照，常年雾气笼罩将强烈的光照折射为茶树生长所需的冷光谱漫射光。茶山毗邻小桥沟国家级自然保护区，森林覆盖率达到87%，年平均降雨量1200毫米，年平均相对湿度湿度达86%。生物多样性极其丰富。茶树大部分种植于原生植被或次生林中，茶地里林木生长茂盛，落叶遍地，腐殖土层深厚，苔藓密布，形成的立体生态非常适合茶树生长。顶层高大乔木形成了遮阴效果，稀疏落下的阳光最有利于高品质茶品的产出，底层的腐殖质和苔藓既保持了水土又为茶树生长提供了丰富的有机质。完备的生态系统无需过多人工干预，病虫害发生率低。

坪寨

以坪寨古茶树鲜叶为原料制作的普洱晒青毛茶，条索肥壮，墨绿油润，干茶香气显著，因其品种持嫩性强的特点，条索较长。当年制作的新茶汤色绿黄明亮；香气高扬，留香持久，前段百花香，有多种花香复合而成，后段略带果香。茶汤滋味平衡，口感刺激度低，苦涩轻微，甜醇润滑，喉韵山野气强，体感绵厚犹如海浪层层推进。压制成饼转化五年后香气蜜韵凸显，带果香及轻微药香，茶汤醇厚浓稠。

（八）麻栗坡

麻栗坡老山古茶树生长环境位于中国三大生物多样性中心的核心部位，多种植被共生，被生物学家定义为生物避难带的"南疆宝地"；气候温暖湿润多云雾；土壤为花岗岩风化后形成的粗粒结晶岩类黄壤，具备适宜茶树生长的良好生态环境。

麻栗坡

麻栗坡老山古茶树种质资源丰富，有古茶园面积3万余亩，达40余万株，品种主要为地方白毛茶变种和普洱茶种等，是当地的优良品种，树龄百年以上且树围超过100厘米的古茶树连片分布，古茶园生物多样性丰富，自然植被良好。

麻栗坡老山雨雾天气多、有效光照足、气候温暖湿润，茶树生长期长，茶叶芽头肥壮、茸毛多、嫩绿光润、甜醇甘爽、滋味醇和、回甘良好、持嫩性好、新梢内含物丰富，适制性广，产品具有"花香高扬、汤甜水滑、回甘生津"的优异品质特征。

（九）勐宋

勐宋乡位于勐海县东部，东与景洪市毗邻，南接格朗和乡，北与勐阿乡相连，西南为勐海镇。乡政府距县城23千米，距景洪39千米，辖曼迈、糯有、曼吕、蚌冈、坝檬、大安、蚌囡、曼方、三迈、曼金10个行政村，111个自然村，102个村民小组。共4781户21467人。主要民族为拉祜族、哈尼族。

勐宋

勐宋，系傣语地名，意为高山上的平坝。山区面积在95%以上，全乡总面积为492.67平方千米，地处横断山脉的南缘地段，地势由西北向东南倾斜，境内山脉大多为南北走向，海拔最高点在西部的滑竹梁子2429米（为全州最高点），最低点在东南部回令河与流沙河交汇处772米，相对高差1657米。海拔1500～2000米以上地区气候温凉，年平均气温16～17℃，年降雨量1500毫米左右。

勐宋古茶山位于勐海县勐宋乡境内，东接景洪市，南连格朗和乡，隔流沙河与南糯山对望。勐宋是勐海最老的古茶区之一，从勐宋保塘村留下的几十亩特大型古茶树来分析，勐宋山区少数民族种茶的历史与南糯山少数民族种茶的历史一样悠久。

勐宋古茶山如今保存下来的古茶园还有3000多亩，主要分布在大安、南本、保塘新寨、保塘旧寨、坝檬、大曼吕、那卡等寨子。保塘离乡政府约10千米，是勐宋乡最具代表性的一个古茶村。勐宋古茶园大多为拉祜族所种，古茶园附近都有拉祜族古寨。清光绪年间已有汉人进入勐宋保塘、南本定居，做茶叶买卖。

勐宋普洱茶条形匀称，纤细秀美、芽毫较显。开汤香气高扬而沉实，汤色黄明透亮，入口苦涩味稍重，尤其涩感较明显，但化得很快，生津一般，口感饱满丰富，回甘强而持久，唯汤质下沉感稍弱。叶底黄润柔韧，光鲜度好。

（十）南糯山

南糯山位于景洪到勐海的公路旁，距勐海县城24千米，是西双版纳有名的茶叶产地。位于东经100°31′~100°39′，北纬21°~22°01′，平均海拔1400米，年降水量在1500~1750毫米，年平均气温16~18℃，十分适宜茶树生长。南糯山村委会辖30个自然村寨，居民均为哈尼族。南糯山被称为茶树王的栽培古茶树，基部径

南糯茶山

围达1.38米，树龄800多年，可惜在1994年死去。在茶树王旁2米左右的地方，现还存活着一株干径超过20厘米的大茶树，据说是茶树王的儿子。后来人们在半坡寨古茶园中新命名了一棵茶王树。

南糯山茶园总面积有21600多亩，其中古茶园12000亩。古茶树主要分布在9个自然村，比较集中的是：竹林寨有茶园2900亩，古茶园1200亩；半坡寨有茶园4200亩，古茶园3700亩；姑娘寨有茶园3500亩；古茶园1500亩。南糯山古茶园由于分布较广不同片区茶的口感滋味有一定区别。

南糯山茶是勐海大叶种的典型代表之一，是江外新六大茶山之一，村村寨寨有古茶，或种于坡地，或与雨林混生，生态环境较好，茶区内品种较多，其中很多优良品种都发源于此，比如大家所熟悉的南糯白毫、云抗系列、紫娟等茶树品种都源

于南糯山。

南糯山古树茶条索较长较紧结，匀称度好，比较显毫，新茶汤色金黄透亮，汤质较饱满；苦味明显，回甘较快，涩味持续时间比苦长，生津较好；新茶香气不扬、不过山野气韵较好、耐泡度较好，叶底黄润，柔韧度好。

（十一）景迈山古茶林

2023年9月17日，中国"普洱景迈山古茶林文化景观"申遗项目在联合国教科文组织第45届世界遗产委员会会议上通过审议，被列入《世界遗产名录》。这是我国第57项世界遗产，云南省第6项世界遗产。

景迈山古茶林位于普洱市澜沧拉祜族自治县，遗产地总面积19095.74公顷，其中

景迈

遗产区面积7167.89公顷，涉及澜沧县惠民镇的景迈、芒景两个村。遗产地居住有傣族、布朗族、哈尼族、佤族、汉族等民族。千百年来，景迈山上布朗族、傣族、哈尼族等各族人民和谐相处，世代与茶共同繁衍、共同发展，形成了丰富多彩的独特民族茶文化，创造了景迈山独特的茶文化景观。

景迈山以"千年万亩古茶林"而闻名，是目前发现成片面积最大、保存最完好、年代最久远的人工栽培型古茶园，有着"世界茶历史文化自然博物馆"的美誉。景迈山拥有古茶林面积2.8万亩，有古茶树320余万株，是传统的"林下茶种植"方式保存至今的实物例证和典型代表，也是见证中国作为茶叶种植起源地的重要实物标识。可以说，景迈山勾勒了茶叶种植的文明史。

景迈山茶树与雨林混生，古茶树成片成林，大多未经修剪矮化，保存完好，现存最大的茶树一株高4.3米，基部干径0.5米，另一株高5.6米，基部干径0.4米。茶园茶树以干径10~30厘米的百年以上老树为主。茶树上寄生有多种寄生植物，其中"螃蟹脚"近年来受到热捧。

茶区内规模化初制较少，基本上是农户自采自制自售，鲜叶采摘比较标准，以一芽二叶为主，嫩度较好。景迈制茶有充分揉捻的传统，条索较紧结，色泽黑亮。景迈茶香气突显、山野之气强烈。由于茶树与森林混生，具有强烈的山野气韵，而且具有特别的、浓郁的、持久的兰花香，被誉为"景迈香"。景迈茶的甜是快速直接的，同时又是持久的。苦弱涩显，耐冲泡。叶底黄绿柔嫩，光鲜度好。

（十二）贺开

贺开村隶属于云南省勐海县勐混镇，地处勐混镇东面，距镇政府所在地8千米，交通方便，距勐海县城20千米。东邻大勐龙镇，南邻格朗河乡，西邻曼蚌村委会，北邻布朗山乡。辖曼贺勐、广冈、曼弄老寨、邦盆老寨等9个村民小组，有农户866户，有乡村人口4060人，从事第一产业人数为2400人。全村面积25.74平方千米，海拔1200米，年平均气温为17.6℃，年降水量1329.6毫米。该村以拉祜族、傣族为主（是拉祜族、傣族混居地），其中拉祜族2259人，傣族1045人，其他民族705人。

贺开

贺开是江外新六大茶山之一，也是云南连片古茶保存面积较大较完整的茶区之一。贺开的古茶树主要分布在曼弄新寨、曼弄老寨、曼迈几个寨子里。茶区内古茶成片成林，多为坡地种植，与雨林混生，无修剪矮化，因前些年交通不便，生态环境较好，2008年后成为新兴古茶山的主要代表。因茶区内古茶树资源丰富，所以近年新建规模化初制所较多，很多一线品牌普洱茶生产商都把贺开当做重要原料基地及茶山旅游重地。贺开古树茶条索黑亮紧结、茶条稍长，油润度好，较显毫，冲泡汤色金黄明亮，有明显苦味，苦化得较快，回甘较快较明显，涩感较明显，化得稍慢，但生津很好，汤质饱满，山野气韵较强，杯底花蜜香明显且较持久，耐泡度好，叶底黄亮柔软。

（十三）忙肺

忙肺茶山位于永德县勐板乡西南边，距离勐板乡8千米，是忙肺村委会所在地。其海拔1500～1600米，年平均气温24℃，年降水量1013毫米。茶区内至今还生长着中华木兰以及大面积的野生型、过渡型、栽培型的古茶树，而忙肺大叶茶属勐库大叶种的引种，现茶区内还有大面积的藤条茶种植。茶区内多是少数民族，又以佤族居多。传统茶园种植管理较为粗犷，茶园多居山坡种植，很少呈梯地状，当地茶园因植被稀疏，故光照充足，茶叶

忙肺

生长状况良好。忙肺茶冲泡后汤色清澈明亮，香气馥郁高扬，口感饱满协调，甘醇顺滑带微涩、舌底生津明显，苦味较重，但回甘快而明显，喉韵甘润持久，叶底柔韧光鲜。

（十四）倚邦

倚邦属古六大茶山之一，地处勐腊县象明彝族乡西北边，距乡政府24千米，距勐腊县198千米。年平均气温25℃，年降水量1700毫米。"倚邦"一词在傣语中的意思是有茶树有水井的地方。倚邦茶区海拔差异大，最高点山神庙1950米，最低点磨者河与小黑江交汇处只有565米，倚邦茶区产茶著名的地方有倚邦、曼松、嶍崆、架布、曼拱等。

倚邦

倚邦茶区种茶历史悠久，明代初期已茶园成片，在曼拱古茶园中还保留着基部径围1.2米，高6米，树龄500年左右的古茶树，至今古茶树保留较多的是麻栗树、倚邦、曼拱等地。

倚邦茶树相较于易武茶树低矮、叶小、芽细、节短，持嫩性差。当地茶农说："比易武茶好，不浑，只要一小点就好，泡多有涩味在。"倚邦本地茶叶以曼松茶叶最好，有吃曼松看倚邦之说。茶树大多山区坡地与雨林混生，无阶梯状，古茶园大多未修剪矮化，采摘较标准，以二叶及一叶居多，基本工艺为鲜叶采摘，适度走水，铁锅高温杀青、轻揉、松条，日晒后茶叶枝梗匀称饱满、芽毫明显、开泡香气迷人，水路细腻，无明显苦涩感，叶底鲜亮油润富活性。

（十五）漭水

漭水镇位于云南省昌宁县中部，距县城16千米，东北部隔澜沧江与临沧地区凤庆县、大理州永平县相望，东南部与本县的达丙、右甸两镇接壤、西部与大田坝乡相连。漭水是一个典型的山区农业镇，辖区土地面积为311平方千米，行政区划分为9个村民委员会，205个村民小组，镇内最低海拔1050米，最高海拔2850米，年平均气温14.5℃，年降雨量为1450毫米左右，有低热河谷气候、温凉和高寒气候等多种气候类型。

漭水茶属昌宁大叶种，种茶历史可追溯到明洪武年间，其历史名茶"碧云仙茶"在历书上有详细记载，漭水茶大面积种植是在20世纪50年代和70年代。当地古茶树资源丰富，除了人工栽培型古茶树外，还有大面积的野生古茶树资源。潞水茶冲泡后汤色黄绿明亮，香气高扬，口感略显单一，有明显涩感、舌底生津明显，苦味较轻，回甘慢但比较持久，耐泡度稍差，叶底柔韧光鲜。

（十六）凤山

凤山镇地处临沧市凤庆县城所在地，是全县的政治、经济和文化中心，也是儒家文化荟萃之地，著名"滇红"茶的发源地。全镇国土总面积218.316平方千米，东与洛党、小湾两镇相接，南与三岔河镇相连，西至勐佑镇、德思里乡，北与大寺乡相邻，境内群山纵横、山峦起伏，最高海拔2863米，最低海拔1472米，森林覆盖率为28%。气候温和，日照充足，雨量集中，干湿分明，冬无严寒，夏无酷暑，年平均气温为16.6℃。雨量丰沛，年平均降雨量1307毫米。全镇辖18个村民委员会，4个社区居民委员会，138个自然村，362个村民小组，居住有汉族、回族、彝族、白族、佤族等20多个民族。

凤山

凤庆大叶种，又名凤庆长叶茶、凤庆种，属于有性系、乔木型、大叶类、早生种。树姿直立或开张；叶形椭圆或长椭圆，叶色绿润，叶面隆起，叶质柔软，便于揉捻成条。嫩芽绿色，满披茸毛，持嫩性强，一芽三叶百芽重9.0克，较勐库大叶种轻，没有勐库种肥壮。凤庆大叶种在1984年被认定为国家级良种，编号为"华茶13号（GsCT13）"。

凤庆种茶制茶历史悠久，现存于小湾镇香竹箐3200年的茶王树被认定为世界上最古老的人工栽培型古茶树，此外明代大旅行家徐霞客游记中所记载的"凤山雀舌""太华茶"等历史名茶均在凤庆境内，凤山茶更是因1938年试制出优质滇红及滇绿茶而闻名。凤山普洱茶冲泡汤色清绿明亮，香气馥郁高扬，特有淡淡篱蒿之清香，口感饱满协调，有明显涩感、舌底生津明显，苦味较轻，回甘较慢但比较持久，叶

潞水

底柔韧光鲜。

二、名茶品鉴

本书中,笔者将近几年品茗过的一些名茶做介绍,带领大家来领悟普洱茶的魅力。

(一)勐海茶厂7572普洱熟茶

勐海茶厂的7572被誉为普洱熟茶的评判标杆。采用200毫升盖碗7克茶冲泡的方法品鉴了20世纪90年代末昆明纯干仓存放的7572普洱熟茶,市场俗称"苹果绿",是精品普洱熟茶中不可多得的稀缺品。

90年代末普洱熟茶外形

1. 外形描述

此茶用料考究,拼配得当,条索肥壮匀整,发酵适度。

2. 香气

香气入汤,平稳持久,清秀显陈香。

90年代末普洱熟茶汤色

3. 汤色和滋味

茶汤红浓明亮,显琥珀色,醇厚甘滑,口齿持久留香,喉韵绵长,生津回甘。

4. 叶底

叶底呈棕褐色,活性较好。

7克茶冲泡10余泡后,依然滋味持久,无水味(茶水分离)。

(二)老班章普洱生茶

采用盖碗冲泡的方法品鉴了2011年和2014年的两款老班章普洱生茶。

90年代末普洱熟茶底

3.汤色和滋味

汤色金黄明亮,茶汤饱满且质滑,回甘生津韵足。

4.叶底

叶底黄绿肥厚,柔韧匀润。

(七)中茶圆梦飞天(典藏版)普洱生茶

采用专业审评杯,邀请评茶师对中茶圆梦飞天(典藏版)普洱生茶进行品鉴。

1. 外形

饼形圆润饱满,松紧适度,条索肥嫩显毫,色泽墨绿油润。

2. 香气

香气蜜香馥郁,陈香优雅透木质香。

3. 汤色和滋味

汤色金黄透亮,滋味浓厚有苦底,回甘强烈,喉韵悠长,气韵相通。

4. 叶底

叶底黄绿肥厚，柔韧匀润。

（八）2022中茶八八青饼（易武版）普洱生茶

采用专业审评杯，邀请评茶师对2022中茶八八青饼（易武版）普洱茶（生茶）紧压茶进行品鉴。

1. 外形描述

饼形圆润，条索紧结，墨绿润泽，芽毫显露。

2. 香气

清香高扬，爽朗悠远，蜜香浓郁，馥郁甘长。

3.汤色和滋味

汤色黄明，清澈透亮，金光潋滟，干净油润；滋味浓醇，饱满厚滑，香甜细腻，韵味长足，耐冲泡。

4.叶底

叶底呈绿黄色，嫩软无杂物，无红梗。

（九）六大茶山贺开有机普洱茶

采用专业审评杯，邀请评茶师对四款不同年份、不同类别的贺开有机普洱茶进行品鉴。

1.外形描述

（1）2019年贺开有机普洱生茶：饼型周正，松紧适度，条索肥壮显毫、紧结匀齐，色泽褐绿油润。

（2）2022年贺开有机普洱生茶：饼型周正，松紧适度，条索肥壮显毫、紧结匀齐，色泽墨绿油润。

（3）2019年贺开有机普洱熟茶：饼型周正，松紧适度，条索紧结匀齐，色泽红褐油润显毫。

（4）2022年贺开有机普洱熟茶：饼型周正，松紧适度，条索紧结匀齐，色泽红褐显毫。

2019年贺开普洱生茶外形　　**2022年贺开普洱生茶外形**　　**2019年贺开普洱熟茶外形**　　**2022年贺开普洱熟茶外形**

2.汤色、滋味和香气

（1）2019年贺开有机普洱生茶：汤色橙黄透亮；滋味醇厚回甘；香气清香浓厚；耐冲泡。

（2）2022年贺开有机普洱生茶：汤色黄明亮；滋味醇厚纯正；香气馥郁浓厚、花蜜香显；耐冲泡。

（3）2019年贺开有机普洱熟茶：汤色红浓透亮；滋味醇厚回甘；香气陈香浓厚、持久；耐冲泡。

（4）2022年贺开有机普洱熟茶：汤色红浓明亮；滋味浓厚回甘；香气馥郁高扬、持久；耐冲泡。

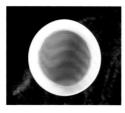   

| 2019年贺开有机普洱生茶汤色 | 2022年贺开有机普洱生茶汤色 | 2019年贺开有机普洱熟茶汤色 | 2022年贺开有机普洱熟茶汤色 |

3. 叶底

（1）2019年贺开有机普洱生茶：叶底绿黄肥壮、匀亮，柔软有弹性，无杂物，无梗。

（2）2022年贺开有机普洱生茶：叶底绿黄匀嫩肥厚，匀亮，柔软无杂物，无梗。

（3）2019年贺开有机普洱熟茶：叶底红褐、匀亮，柔软有弹性，无杂物，稍有梗。

（4）2022年贺开有机普洱熟茶：叶底红褐、匀亮，柔软有弹性，无杂物，稍有梗。

| 2019年贺开有机普洱生茶叶底 | 2022年贺开有机普洱生茶叶底 | 2019年贺开有机普洱熟茶叶底 | 2022年贺开有机普洱熟茶叶底 |

（十）德凤官寨普洱生茶

采用盖碗冲泡方法，邀请评茶师对两款不同年份的德凤官寨普洱生茶进行品鉴。

1. 外形描述

（1）2015年德凤官寨普洱生茶：饼型规整匀齐，条索肥硕、紧结完整，油亮多毫。

2015年德凤官寨普洱生茶外形

（2）2019年德凤官寨普洱生茶：饼型规整匀齐，条索肥硕、紧结完整，绿黄显毫。

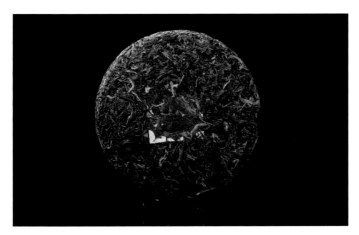

2019年德凤官寨普洱生茶外形

2. 香气

（1）2015年德凤官寨普洱生茶：果香馥郁透木香，通透、干净、具有穿透力。

（2）2019年德凤官寨普洱生茶：茶汤、杯底香气高度统一；香气浓郁，花果蜜香，清扬持久。

3. 汤色和滋味

（1）2015年德凤官寨普洱生茶：汤色金黄为主，明亮度较高，轻摇有油质感；入口香韵升起，水路细腻，舌面甘甜顺滑，浓厚甘醇回甘持久，喉韵佳，茶气强烈，喉部有回甘和留香；耐冲泡。

2015年德凤官寨普洱生茶汤色

（2）2019年德凤官寨普洱生茶：汤色浅黄透亮，汤感黏稠；入口甜润，随之有淡淡苦涩，又迅速化开，涩不挂口，苦不留舌，饮后整个口腔至喉咙清爽甘纯，舌面不时泛起丝丝甜韵；杯香隐隐，细柔绵长香气挂于杯壁。

2019年德凤官寨普洱生茶汤色

4. 叶底

（1）2015年德凤官寨普洱生茶：叶质柔软油润，活性十足，富有弹性；油润度高。

2015 年德凤官寨普洱生茶叶底

（2）2019年德凤官寨普洱生茶：叶底柔软无杂物，无红梗。

2019 年德凤官寨普洱生茶叶底

（十一）文山坪寨古树普洱生茶

采用专业审评杯，邀请评茶师对坪寨古树普洱生茶进行品鉴。

1. 外形描述

肥硕润亮，显毫，透木香；具有"肥亮显毫"的特点。

2.香气

花蜜香浓郁、高长、持久,富有穿透力;具有"花香兰韵"的特色。

3.汤色和滋味

汤色金黄、透亮、稠厚;具有"透亮稠厚"的特征;滋味柔甜、厚润、甘爽,木质甜饱满丰富,经久耐泡;具有"柔甜厚润"的特点。

4.叶底

肥嫩匀亮、活性好;具有"肥嫩鲜活"的特征。

(十二)科技普洱·坪寨古树普洱熟茶

采用专业审评杯,邀请评茶师对科技普洱·坪寨古树普洱熟茶进行品鉴。

1.外形描述

饼形周正、棕褐润亮、显金毫。

2.香气

馥郁悠长、具有陈香、药香、木香、曲酯香、菌香、杏仁香。

3. 汤色和滋味

汤色红浓明亮,滋味甜醇回甘、稠滑、顺滑。

4. 叶底

棕褐软亮活性好。

参考文献：

[1] 王白娟，张贵景. 云南普洱茶的饮用与品鉴[M].昆明:云南科技出版社,2015.

[2] 宛晓春.中国茶谱(第2版)[M].北京:中国林业出版社,2010.

[3] 宛晓春,夏涛,等.茶树次生代谢[M].北京:科学出版社,2015.

[4] 陈椽.茶叶通史[M].北京:中国农业出版社,2008.

[5] 安徽农学院.制茶学(第2版)[M].北京:中国农业出版社,2012.

[6] 宛晓春,李大祥,张正竹,等.茶叶生物化学研究进展[J].茶叶科学,2015,35(1):1-10.

[7] 陈椽.茶叶分类的理论与实际[J].茶叶通报,1979(Z1):48-56,94.

[8] 李大祥,王华,白蕊,等.茶红素化学及生物学活性研究进展[J].茶叶科学,2013,33(4):327-335.

[9] 宋丽,丁以寿. 陈椽茶叶分类理论[J].茶叶通报,2009,31(3): 143-144.

[10] 刘勤晋.茶文化学[M].中国农业出版社,2000.

[11] 王玲.中国茶文化[M].中国书店,1998.

[12] 杨崇仁,陈可可,张颖君.茶叶的分类与普洱茶的定义[J].茶叶科学技术,2006(2):37-38.

[13] 熊昌云.普洱茶降脂减肥功效及作用机理研究[D].杭州:浙江大学,2012.

[14] Després JP, Lemieux I. Abdominal obesity and metabolic syndrome [J]. Nature,2006,444(7121): 881-887.

[15] Houstis N, Rosen E D, Lander E S. Reactive oxygen species have a causal role in multiple forms of insulin resistance [J]. Nature,2006, 440(7086):944-948.

[16] Rosen E D, Spiegelman B M. Adipocytes as regulators of energy balance andglucose homeostasis[J].Nature,2006, 444(7121):847-853.

[17] 吴文华.晒青毛茶、普洱茶降血脂功能比较[J].福建茶叶,2004(4):30.

[18] 周红杰,秘鸣,韩俊,等.普洱茶的功效及品质形成机理研究进展[J].茶叶,2003,29(2):75-77.

[19] 东方.普洱茶的抗氧化特性及活性成分鉴定[D].杭州:浙江大学,2007,5.

[20] 折改梅,张香兰,陈可可,等.茶氨酸和没食子酸在普洱茶中的含量变化[J].云南植物研究,2005,27(5):572-576.

[21] Floyd R A,Towner R A,He T,et al. Translational research involving oxidative

stress and diseases of aging [J]. Free Radical Biology and Medicine., 2011, 51(5): 931–941.

[22] Liu C C, Gebicki J M. Intracellular GSH and ascorbate inhibit radical-induced protein chain peroxidation in HL-60 cells [J]. Free Radic. Biology and Medicine, 2012, 52(2): 420–426.

[23] Niki E. Assessment of Antioxidant Capacity in vitro and in vivo [J]. Free Radic. Biology and Medicine, 2010, 49(4): 503–515.

[24] 朱旗,Clifford M N,毛清黎,等.LC-MS分析普洱茶和茯砖茶与红茶成分的比较研究[J].茶叶科学,2006,26(3):191–194.

[25] 凌关庭.抗氧化食品与健康[M].北京:化学工业出版社.2004.

[26] 金裕范,高雪岩,王文全,等.云南普洱茶抗氧化活性的比较研究[J].中国现代中药.2011,13(8):17–19.

[27] 江新凤,邵宛芳,侯艳.普洱茶预防高血脂症及抗氧化作用的研究[J].云南农业大学学报.2009,24(5):705–711.

[28] 任洪涛,周斌,秦太峰,等.普洱茶挥发性成分抗氧化活性研究[J].茶叶科学.2014,34(3):213–220.

[29] 陈浩.普洱茶多糖降血糖及抗氧化作用研究[M].杭州:浙江大学,2013.

[30] 林瑞萱.中日韩英四国茶道[M].北京:中华书局,2008.

[31] 吴远之.大学茶道教程[M].北京:知识产权出版社,2011.

[32] 高力,刘通讯.不同储藏时间的普洱茶内所含成分及其抗氧化性质研究[J].食品工业,2013,34(7):127–130.

[33] 李银梅.普洱茶在不同贮藏条件下品质及成分变化初探[J].茶叶通报,2010,32(1):46–48.

[34] 周黎,赵振军,刘勤晋,等.不同贮藏年份普洱茶非挥发物质的GC-MS分析[J].西南大学学报,2009,31(11):140–144.

[35] 邓时海.普洱茶[M].昆明:云南科技出版社,2004.

[36] 周红杰.云南普洱茶[M].昆明:云南科技出版社,2004.

[37] 熊志惠.识茶、泡茶、鉴茶全图解[M].上海:上海科学普及出版社.2011.

(3) Tea liquor color and taste

The liquor color is red, bright and deep, its taste is mellow, smooth, fresh, and it has a sweet aftertaste.

(4) Infused tea leaves

The infused tea leaves are plump and tender with brownish auburn, and even with vitality.

(Refer to the Chinese references for the English references.)

(3) Tea liquor color and taste

The liquor color is golden yellow and bright and deep. The taste is sweet, mellow and thick. It is full of woody sweet, and it can be infused for several times. It has the characteristic of being "soft, sweet, thick and moist".

(4) Infused tea leaves

The infused tea leaves are fleshy, tender and even, with good activity. They have the characteristic of being "plump, tender, and lively".

2.12 Keji *Pu'er* - Pingzhai Ancient Tea Tree Ripe *Pu'er* Tea

Adopt professional tea tasting cups and invite tea tasters to taste Pingzhai Ancient Tree ripe *Pu'er* Tea.

(1) Shape

The tea cake is flat and even with blooming brownish auburn, and has golden pekoe.

(2) Aroma

This tea has a fragrant and lasting aroma, including a aroma after aging, a herb aroma, a woody aroma, a distiller's yeast aroma, a mushroom aroma, a almond aroma and so on.

b) Defeng Guanzhai raw *Pu'er* tea in 2019 : The leaves are soft and free of impurities, and there are no red stalks after brewing.

Infused tea leaves of Defeng Guanzhai Raw *Pu'er* tea in 2019

2.11 Wenshan Pingzhai Ancient Tea Tree Raw *Pu'er* Tea

Adopt professional tea tasting cups, and invite tea tasters to taste Pingzhai Ancient Tree raw *Pu'er* Tea.

(1) Shape

The tea cake is plump and bright, and slightly tippy with a woody aroma.

(2) Aroma

The tea has a flowery and honey aroma, which is strong, intensive and lasting, and has the characteristic of a magnolia aroma.

Tea liquor color of Defeng Guanzhai Raw *Pu'er* tea in 2015

b) Defeng Guanzhai raw *Pu'er* tea in 2019 : The liquor is light yellow and transparent. The taste is sweet followed by a slight bitter. After drinking, the whole mouth to the throat is refreshing and sweet. A soft and long aroma remains in the cup.

Tea liquor color of Defeng Guanzhai Raw *Pu'er* tea in 2019

(4) Infused tea leaves

a) Defeng Guanzhai raw *Pu'er* tea in 2015 : The leaves are soft and blooming, full of vitality and elasticity and rich in oil content.

Infused tea leaves of Defeng Guanzhai Raw *Pu'er* tea in 2015

Shape of Defeng Guanzhai Raw *Pu'er* Tea in 2015

b) Defeng Guanzhai Raw *Pu'er* Tea in 2019: Regular and even in shape, plump and fat in strips, tight and complete in texture, greenish yellow and slightly tippy.

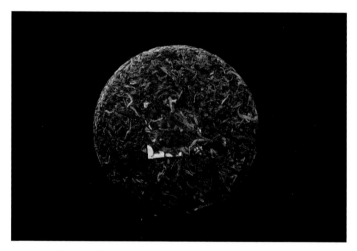

Shape of Defeng Guanzhai Raw *Pu'er* Tea in 2019

(2) Aroma

a) Defeng Guanzhai Raw *Pu'er* Tea in 2015: Fragrant and lasting, fruity aroma with woody aroma, transparent, clean and penetrating.

b) Defeng Guanzhai Raw *Pu'er* Tea in 2019: The aroma of tea liquor and cup is highly unified, which is strong and lasting, along with a flowery and fruity honey aroma.

(3) Tea liquor color and taste

a) Defeng Guanzhai raw *Pu'er* tea in 2015 : The liquor is golden yellow, bright, and with an oil texture when softly shaken. The taste is sweet and smooth, and it has a lasting sweet aftertaste. More infusing times.

Tea Liquor color of HeKai Organic Raw *Pu'er* Tea in 2019, 2022

Tea liquor color of HeKai Organic Ripe *Pu'er* Tea in 2019, 2022

(3) Infused tea leaves

a) Hekai organic tea in 2019

Pu'er tea (raw tea): Infused tea leaves are greenish, yellow, plump, evenly bright, soft and elastic, with no impurities or stalks.

b) Hekai organic tea in 2022

Pu'er tea (raw tea): Infused tea leaves are greenish yellow, tender, thick, evenly bright, soft, with no impurities or stalks.

c) Hekai organic tea in 2019

Pu'er tea (ripe tea): Infused tea leaves are reddish auburn, evenly bright, soft and elastic, with no impurities but slight stalks.

d) Hekai organic tea in 2022

Pu'er tea (ripe tea): Infused tea leaves are reddish auburn, evenly bright, soft and elastic, with no impurities but slight stalks.

Infused tea leaves of HeKai Organic Raw *Pu'er* Tea in 2019, 2022

Infused tea leaves of HeKai Organic Ripe *Pu'er* Tea in 2019, 2022

2.10 Defeng Guanzhai Ancient Tea Tree Raw *Pu'er* Tea

Tea tasters are invited to taste two different types of raw *Pu'er* tea of ancient tea trees of different years in Defeng Guanzhai by the method of lidded-tea bowl brewing.

(1) Shape

a) Defeng Guanzhai Raw *Pu'er* Tea in 2015: Regular and evenshape, plump and fat in strips, tight and complete texture, shiny and fairly tippy.

b) Hekai organic tea in 2022

Pu'er tea (raw tea): The cake-shaped tea is regular and moderately tight. Tea strips are plump, even, slightly tippy, dark green and blooming.

c) Hekai organic tea in 2019

Pu'er tea (raw tea): The cake-shaped tea is regular and moderately tight. Tea strips are plump, even, slightly tippy, reddish auburn and blooming.

d) Hekai organic tea in 2022

Pu'er tea (raw tea): The cake-shaped tea is regular and moderately tight. Tea strips are plump, even, slightly tippy and reddish auburn.

Shape of HeKai Raw *Pu'er* Tea in 2019, 2022   **Shape of HeKai Ripe *Pu'er* Tea in 2019, 2022**

(2) Tea liquor color, taste, and aroma

a) Hekai organic tea in 2019

Pu'er tea (raw tea): The liquor has a bright orange color. The taste is mellow and sweet. The aroma is clean and refreshing. More infusing times.

b) Hekai organic tea in 2022

Pu'er tea (raw tea): The liquor has a bright yellow color. The taste is mellow and pure. The smell is fragrant and lasting, with a flowery and honey aroma. More infusing times.

c) Hekai organic tea in 2019

Pu'er tea (ripe tea): The liquor has a rich and transparent red color. The taste is mellow and sweet. The aroma is rich and lasting. More infusing times.

d) Hekai organic tea in 2022

Pu'er tea (ripe tea): The liquor has a thick and bright red color. The taste is strong and sweet. The aroma is fragrant and lasting. More infusing times.

(2) Aroma

The smell is refreshing and lingering, with a rich and sweet honey aroma.

3. Tea liquor color and taste

The liquor is yellow, clear and transparent, shining with golden light, clean and blooming. The taste is rich and mellow, full and smooth, sweet and delicate, with a long-lasting lingering charm and more infusing times.

4. Infused tea leaves

The infused tea leaves are greenish yellow, tender and soft with no impurities or red stalks.

2.9 Hekai Organic *Pu'er* Tea Produced by Yunnan Six Tea Mountain Industry CO., Ltd

Tea tasters are invited to taste four different types of Hekai organic teas from different years.

(1) Shape

a) Hekai organic tea in 2019

Pu'er tea (raw tea): The cake-shaped tea is regular and moderately tight. Tea strips are plump, even, slightly tippy, auburnish green and blooming.

(2) Aroma

The smell is with a honey aroma, fragrant and lasting; with an elegant aroma after aging and woody aroma.

(3) Tea liquor color and taste

The liquor has a golden and transparent color, a strong and bitter taste, a strong and long-lasting aftertaste, and a consistent flavor.

(4) Infused tea leaves

The Infused tea leaves are yellowish green and plump, soft and even.

2.8 2022 Zhongcha "Ba Ba Green Tea" (Yiwu version) Raw *Pu'er* Tea

Tea tasters are invited to use professional tasting cups to taste this tea.

(1) Shape

The shape is round and full, dark green and blooming with tightly knotted strips and exposed buds.

(2) Aroma

The tea is fragrant, rich in flavor, mellow and elegant, fleshy and plump, powerful and distinctive.

(3) Tea liquor color and taste

The liquor is of a bright golden color, with a full and smooth flavor, refreshing and enriching saliva, and a sweet aftertaste.

(4) Infused tea leaves

The infused tea leaves, of a yellowish green color, are fleshy and plump, soft and even.

2.7 Zhongcha "Dream of Flying High" (collection version) Raw *Pu'er* Tea

Professional tasters are invited to taste this tea with professional tasting cups.

(1) Shape

The shape is round and full, with moderate elasticity, plump and slightly tippy, dark green and blooming.

(2) Aroma

The aroma is intense with no dull odour, full and mellow.

(3) Tea liquor color and taste

The liquor is red with glossy, thick and clearly bright, looking like red wine. The taste is thick and smooth, with a strong and wild flavor.

(4) Infused tea leaves

The infused tea leaves are reddish brown and blooming, with full vitality.

2.6 Zhenzi Hao Longzhang Raw *Pu'er* Tea

Adopting professional tea tasting cup, and inviting tea tasters to taste Longzhang Raw *Pu'er* tea.

(1) Shape

The tea cake is round and blooming, with tightly twisted strips, dark green and moist, and the buds are completely exposed.

Tea liquor color of Shuangjiang Mengku Ancient Arbor Tea in 2005, 2008, 2011, 2015

(3) Infused tea leaves

Ancient Arbor Tea in 2015: Infused leaves are greenish yellow, tender and soft, with no impurities or red stalks.

Ancient Arbor Tea in 2011: Infused leaves are yellowish green, even, with no impurities or red stalks.

Ancient Arbor Tea in 2008: Infused leaves are yellow, soft and extensible, with no impurities but slightly red stalks.

Ancient Arbor Tea in 2005: Infused leaves are golden yellow, soft and flexible, with no impurities but red stalks.

Infused tea leaves of Shuangjiang Mengku Ancient Arbor Tea in 2005, 2008, 2011, 2015

2.5 Zhenzi Hao Longxi Ripe *Pu'er* Tea

Adopt professional tea tasting cups, and invite tea tasters to taste Longxi Ripe *Pu'er* tea.

(1) Shape

The tea cake is round and blooming, with tightly twisted strips, dark green and moist, and the buds are completely exposed.

2.4 Shuangjiang Mengku Ancient Arbor Tea

Tea tasters are invited to taste four varieties of Shuangjiang Mengku Ancient Arbor Tea of different years with a professional tasting cup.

(1) Shape

Ancient Arbor Tea in 2015: The cake-shaped tea is regular and even. Tea strips are fat, fleshy, tight, greenish yellow and slightly tippy.

Ancient Arbor Tea in 2011: The cake-shaped tea is regular and even. Tea strips are fat, fleshy, tight, green and bloom and slightly tippy.

Ancient Arbor Tea in 2008: The cake-shaped tea is regular and even. Tea strips are fat, fleshy, tight, yellowish auburn and bloom.

Ancient Arbor Tea in 2005: The cake-shaped tea is regular and even. Tea strips are fat, fleshy, tight, auburnish black and blooming.

Shape of Shuangjiang Mengku Ancient Arbor tea in 2005,2008,2011,2015

(2) Tea liquor color, taste and aroma

Ancient Arbor Tea in 2015: The liquor is greenish yellow and transparent. The taste is fresh, sweet and a little bitter. The aroma is full, flowery. More infusing times.

Ancient Arbor Tea in 2011: The liquor is yellowish green and transparent. The taste is sweet, smooth and slightly bitter. The aroma is strong and lasting, honey with a flowery aroma. More infusing times.

Ancient Arbor Tea in 2008: The liquor is yellow and transparent. The taste is mellow, smooth and slightly bitter. The smell is fragrant and lasting, flowery and fruity with a honey smell. More infusing times.

Ancient Arbor Tea in 2005: The liquor is golden yellow and transparent. The taste is smooth, refreshing and enriching the saliva, with no bitter. The smell is fragrant and lasting, with a honey and fruity aroma. More infusing times.

Bingdao Ancient Tea Tree Raw *Pu'er* Tea in 2015: Strong flowery and honey aroma, no odor.

(3) Cold-cup aroma

Bingdao Ancient Tea Tree Raw *Pu'er* Tea in 2012: Intense and lasting rock candy flavor.

Bingdao Ancient Tea Tree Raw *Pu'er* Tea in 2015: Intense and lasting rock candy flavor.

(4) Tea liquor color and taste

Bingdao Ancient Tea Tree Raw *Pu'er* Tea in 2012: Golden, bright and smooth liquor without bitterness but with a lasting aftertaste.

Bingdao Ancient Tea Tree Raw *Pu'er* Tea in 2015: Bright, yellow liquor with a rock candy flavor.

Tea liquor color of Bingdao Ancient Tea Tree Raw *Pu'er* Tea in 2012

Tea liquor color of Bingdao Ancient Tea Tree Raw *Pu'er* Tea in 2015

(5) Infused tea leaves

Bingdao Ancient Tea Tree Raw *Pu'er* Tea in 2012: The liquor is golden yellow and bright. The taste is smooth with no bitter, and it has a sweet aftertaste, refreshing and enriching the saliva, with a clear rock candy flavor.

Bingdao Ancient Tea Tree Raw *Pu'er* Tea in 2015: The liquor is yellow and bright. The taste is full, with no bitter, and it has a sweet aftertaste, refreshing and enriching the saliva, with a rock candy flavor.

Infused tea leaves of Bingdao Ancient Tea Tree Raw *Pu'er* Tea in 2012

Infused tea leaves of Bingdao Ancient Tea Tree Raw *Pu'er* Tea in 2015

(5) Infused tea leaves

Laobanzhang Raw *Pu'er* tea in 2011: Strong and lasting flowery and fruity aroma, no stale odor. Leaves are soft and bright, with no coarse impurities and no burnt flakes, but slightly red stalks.

Laobanzhang Raw *Pu'er* tea in 2014: Intense flowery and fruity aroma, slightly honey, no odor. Leaves are yellow, bright, fleshy, free of coarse impurities and burnt flakes, with no red stalks.

Infused tea leaves of Laobanzhang *Pu'er* Tea in 2011 and 2014

2.3 Bingdao Ancient Tea Tree Raw *Pu'er* Tea

Two varieties of Bingdao Ancient Tea Tree Raw *Pu'er* tea stored in 2012 and in 2015 are tasted with lidded-tea bowls.

(1) Shape

Bingdao Ancient Tea Tree Raw *Pu'er* Tea in 2012: The cake-shaped tea is regular, even and plump. Tea strips are yellow-bloomed.

Bingdao Ancient Tea Tree Raw *Pu'er* Tea in 2015: The cake-shaped tea is regular, even and plump. Tea strips are blooming and shiny.

Shape of Bingdao Ancient Tea Tree Raw *Pu'er* Tea in 2012   **Shape of Bingdao Ancient Tea Tree Raw *Pu'er* Tea in 2015**

(2) Warm-cup aroma

Bingdao Ancient Tea Tree Raw *Pu'er* Tea in 2012: Ripe fruity aroma, no stale odor.

(1) Shape

Laobanzhang raw *Pu'er* tea in 2011: The compressed cake-shaped tea is regular, even and plump. Tea strips are yellow-bloomed, fat and fresh.

Laobanzhang *Pu'er* tea in 2014: The compressed cake-shaped tea is regular, even and plump. Tea strips are fat, full, blooming and slightly tippy.

Shape of Laobanzhang *Pu'er* tea in 2011   **Shape of Laobanzhang *Pu'er* tea in 2014**

(2) Warm-cup aroma

Laobanzhang raw *Pu'er* tea in 2011: Light flowery aroma, slightly fruity aroma, no stale odor.

Laobanzhang raw *Pu'er* tea in 2014: Clear flowery aroma, no odor.

(3) Cold-cup aroma

Laobanzhang raw *Pu'er* tea in 2011: Honey aroma is intense and lasting.

Laobanzhang raw *Pu'er* tea in 2014: Honey aroma is intense and lasting.

(4) Tea liquor color and taste

Laobanzhang raw *Pu'er* tea in 2011: The liquor is golden yellow and bright. The taste is smooth and full, with a fruity aroma but no bitter, and it has a sweet aftertaste, refreshing and enriching the saliva.

Laobanzhang raw *Pu'er* tea in 2014: The liquor is yellow and bright. The taste is smooth with a flowery aroma, with no bitter but slight astringency;it has a lasting sweet aftertaste, refreshing and enriching the saliva.

Tea liquor color of Laobanzhang   **Tea liquor color of Laobanzhang**
Raw *Pu'er* tea in 2011   **Raw *Pu'er* tea in 2014**

refreshing saliva at the bottom of our tongue. In addition to a light bitterness in taste, the recovery of sweetness is slow and takes a relatively long period. When presented, infused leaves are flexible and bright.

2. Famous *Pu'er* Tea Brands in Yunnan Province

To further understand the charm of *Pu'er* tea, this part will give a brief introduction to the 12 famous *Pu'er* tea brands in Yunnan Province.

2.1 7572 Ripe *Pu'er* Tea by Menghai Tea Factory

7572 ripe *Pu'er* tea has been praised as a benchmark for ripe *Pu'er* tea. This book uses the method of brewing seven grams tea in 200cc lidded bowl to taste the 7572 ripe *Pu'er* tea stored in the pure dry warehouse in Kunming in the late 1990s, commonly nicknamed and known as "Apple Green" in the market, which is a rare product among the fine ripe *Pu'er* teas.

The shape of ripe *Pu'er* tea in the late 1990s

(1) Shape

This tea is made of exquisite raw materials, properly blended, fat and even, and moderately fermented.

Tea liquor color of ripe *Pu'er* tea in the late 1990s

(2) Aroma

Fragrant, lasting, and delicate, with an aroma after aging.

(3) Tea liquor color and taste

Tea liquor is rich and bright, with an amber color, a mellow and smooth, persistent aroma of mouth, and a sweet aftertaste, refreshing and enriching the saliva.

Infused tea leaves of ripe *Pu'er* tea in the late 1990s

(4) Infused leaves

Infused leaves are greenish brown with vitality. Seven grams of tea is used to brew for more than ten times, but it still tastes good.

2.2 Laobanzhang Raw *Pu'er* Tea

Two kinds of Laobanzhang raw *Pu'er* Tea from 2011 and 2014 are tasted with lidded bowls.

detail. The large-scale planting exercises unfolded in the 1950s and 1970s since the local ancient tea resources were rich in measurement. In addition to the artificially cultivated ancient tea, there are also a large area of wild ancient tea resources. Lightly roll, dry in sunlight until the oily green and shiny appearance are revealed and buds are hidden. The color of the tea is bright yellow and green, the aroma is touching, but the taste is slightly monotonous coupled with obvious astringency. More importantly, the saliva on the bottom of tongue is resumed after the slight bitterness, and the recovery of sweetness is slow but endures a relatively long period. Though the infusing times are barely satisfactory, the infused leaves are flexible and bright.

1.16 Tea Tree Planting in Fengshan Town

Fengshan Town is the political, economic and cultural center of the county fengqing, also serves the place where Confucian culture gathers and is the birthplace of the resounded Yunnan black tea. With a total land area of 218.316 square kilometers, the Town manages to connect Luodang and Xiaowan towns in the east, Sanchahe Town in the south, Mengyou Town and Desili Town in the west, and Dasi Town in the north. The mountains are crisscross and undulating within the Town, with the highest altitude of 2,863 meters and the lowest altitude of 1,472 meters. The forest coverage rate is up to 28% and the climate is on the whole mild as we can feel, with sufficient sunshine, concentrated rainfall, distinct dryness and wetness. There are no hash cold days in winter and no severe heat in summer. Besides, the annual average temperature is

Fengshan

16.6℃, with an average annual rainfall of 1,307 millimeters. The Town has jurisdiction over 18 villagers' committees, 4 community residents' committees, 138 natural villages and 362 villagers' groups, which are inhabited by more than 20 nationalities, including people of Han, Hui, Yi, Bai and Wa.

Fengqing large-leaved tea, also known as Fengqing long-leaved tea, belongs to arbor type, large-leaved class and early-growing species. The trees are upright and the leaves are oval with green color and soft texture, which makes it easy to be rolled into strips. The buds are green, covered with pubescence, and have strong tenderness. One bud is coupled with three leaves and surprisingly hundreds of tea buds only weigh about 9.0 g in total, which is lighter than that of Mengku large-leavesd species, but is not as thick as Mengku species. It was first identified as an upgraded variety in China in 1984, and its set number was "GsCT13".

Fengqing has a long history of growing and processing tea. The king tree of tea, which is extant in Xiaowan Town for 3,200 years, is recognized as the oldest artificially cultivated ancient tea tree in the world. The historically famous teas were recorded by Xu Xiake, a well-known traveler in the Ming Dynasty, are all in the Town, which got famous for producing black and green tea of high-quality in 1938. When we take a sip of the Fengshan tea, it's easy to be aware that the tea liquor is bright in color, and holds a strong aroma with a faint sense of artemisia. The taste is full and coordinated, exerting an obvious astringent flavor and producing

Xikong, Jiabu, Mangong, etc. Besides, the tea area has a long history of growing tea and has been developed as a tea garden as early as in Ming Dynasty. In Mangong Ancient Tea Garden, there are still ancient tea trees with a base diameter of 1.2 meters, a height of 6 meters, and trees aging about 500 years. Up to now, most of the ancient tea trees are remained as the old days in places such as Malishu, Yibang, and Mangong.

Overall, Yibang tea trees are lower than Yiwu ones, with smaller leaves, thinner buds, shorter nodes and poorer tenderness. The farmers said, "Yibang tea enjoys better reputation than Yiwu tea does, since it is not muddy when soaked. When we decide to make the tea, only with a little bit of tea blended in liquor, and the enjoyment could be quite considerable, while overdose of tea could be marring the taste to an annoying astringent degree."

In this area, most of the tea species indicate themselves the type of small-leaved tea species. Mansong tea ranks the best local tea in Yibang, besides, there boosts a saying that when you want to take a closer look at local tea in Mansong, Yibang tea is never ought to be overlooked. Also, most tea trees here are mixed with rain forests on mountainous slopes, where most tree of the ancient tea gardens have not been pruned or dwarfed, and are not arbor-shaped. The plucking of tea is relatively on a standard basis, and most of the tea plucking materials are displayed with two leaves and one leaf in a large one. There is no, if any, large-scale primary manufacture in the tea area, which is replaced by a mode of self-functioning by human force. The basic process involves fresh leaves plucking, dewatering, fixing in an iron pot at high-temperature, gently rubbing and loosening the strips. By doing so, the branches are symmetrical and full, the buds are clear, the aroma is charming, the ingress of sipping is exquisite, and there is no obvious bitterness that our taste buds can take in. The infused leaves are bright in color, oily and compact in texture, and basically free of red stems.

1.15 Tea Tree Planting in Mangshui Town

Mangshui Town is located in the middle of Changning County, Yunnan Province, 16 kilometers away from the county seat, facing Fengqing County of Lincang City and Yongping County of Dali Prefecture, across the Lancang River in the northeast, borders Dabing and Youdian towns of the County in the southeast, and connects Datianba Township in the west. Mangshui is a typical agriculture-based town in the mountainous area, with a land area of 311 square kilometers, and administrative divisions of 9 villagers' committees and 205 villagers' groups. In geography, the lowest altitude in the town is 1,050 meters, while the highest altitude is 2,850 meters. In climate, the annual average temperature is 14.5 ℃, and the annual rainfall is about 1,450 millimeters.

There are many kinds of climate types, with Mangshui tea belonging to eternal-green large-leaved tea species. The history of tea planting can be traced back to the Hongwu Period of the Ming Dynasty, with "Tea in Eternal Green" being recorded in

Mangshui

production centers in recent years. Many first-line brand *Pu'er* tea manufacturers regard Hekai as an important raw material base and an important tourist destination of tea mountain as well. Hekai Ancient Tree Tea is dark in color, bright and tight in texture, with slightly longer tea strips and moistening degree in all traits. When we take a sip of the tea, we can get a feeling of obvious bitterness, then speedy recovery of sweetness can easily smarten up the process of enjoyment. The tea liquor in strained with obvious astringency, and slightly slow melting, but it produces good sense of fluid, strong wild flavor, obvious and lasting nectar aroma at the bottom of the cup. The process will finally exert more infusing times.

1.13 Tea Tree Planting in Mangfei Village

Located in the southwest of Mengban Township, Yongde County, Mangfei Tea Mountain is 8 kilometers away from Mengban Township. It is the site of the village committee of Mangfei. The altitude is 1,500-1,600 meters, with the annual average temperature 24℃, and the annual precipitation 1,013 millimeters. Up to now, *Magnolia sinensis* and a large area of wild and transitional cultivated ancient tea trees still can be seen in the tea area, while large-leaved tea is a kind of national tea and also a branch of Mengku large tea. We can notice a large area of rattan tea plantings in the current tea area where most of

Mangfei

them are marked with traits of ethnic minorities with Wa people being the majority. Therefore, the traditional tea plantation management is relatively on a rough basis, in which most of the tea plantations are set on hillsides, rarely in terraces. Because of the lack of vegetation, the local tea plantations can get sufficient sunlight and the tea grows in a fairly sound condition.

When we take a sip of the tea, the tea strips are pale, green and shiny with the buds. Furthermore, the color of tea is clear and bright, the aroma is rich and strong, the taste is full and coordinated, the glycerine is smooth along with slight astringence, and finally the feeling of saliva on the bottom of tongue is quite obvious. Though bitter taste is heavily carried during the sipping process, the sweet taste is on its recovery rather quickly. Overall, the infused leaves are flexible in texture and bright in color, the fixing is even and red stems are hard to be spotted .

1.14 Tea Tree Planting in Yibang Mountain

As one of the six ancient tea mountains, Yibang is located in the northwest of Xiangming Township, Mengla County, 24 kilometers away from the township government and 198 kilometers away from Mengla County, with the annual average temperature 25℃, and the annual precipitation 1,700 millimeters. "Yibang" means a place with abundant teas and wells in Dai language. The altitude of ancient Yibang tea area varies greatly. The highest point is the mountain temple, which stands 1,950 meters above sea level, and the lowest point is the intersection of Mozhe River and Xiaoheijiang River, which is only 565 meters

Yibang

above sea level. The famous places for tea production in Yibang tea area include Yibang, Mansong,

of the traditional "under-story growing tea" method preserved to date, and is also an important physical identification witnessing China as the origin of tea planting. It can be said that Jingmai Mountain outlines the civilization history of tea planting.

The tea trees in Jingmai Mountain are mixed with the rain forest, old tea trees into pieces of forests, mostly without pruning dwarf, well-preserved. The two largest existing tea trees, one of which is 4.3 meters high with a basal stem diameter of 0.5 meters, and the other is 5.6 meters tall with a basal stem diameter of 0.4 meters. The tea trees in the tea garden are mainly over 100 years old trees with a trunk diameter of 10-30 centimeters. There are many kinds of parasitic plants on the tea trees, among which "Viscum liquorambaricolum Hayata" has been popular in recent years.

There are fewer large-scale primary production processes in the tea area, and tea is basically self-harvested, self-produced and self-sold by farmers. There is a clear definition of the standards for fresh leaf harvesting, with one bud and two leaves as the main focus, so that the tenderness is just right. Jingmai tea has the tradition of full twisting and kneading, with tight tea leaves and dark color. The aroma of Jingmai tea is prominent and the atmosphere of the mountains is strong. Due to the tea trees and forests mixed, it not only has a strong mountain atmosphere, but also has a special, rich, long-lasting orchid flavor, known as the unique "Jingmai incense". The sweetness of Jingmai tea is quick and direct, and at the same time, long-lasting. The bitterness and astringency are weak and resistant to brewing. The infused leaf is yellowish green and soft, with good brightness.

1.12 Tea Tree Planting in Hekai Village

Hekai Village is subordinate to Menghun Town, Menghai County, Yunnan Province. It is located in the east of Menghun Town, 8 kilometers away from the town government and 20 kilometers away from Menghai County. Though the road to the township is untidy, the transportation is easy to get access to. It is adjacent to Damenglong Town in the east, Gelanghe Township in the south, Manbang Village Committee in the west and Bulangshan Township in the north. It obtains jurisdiction

Hekai

over 9 villagers' groups, with 866 households, 4,060 rural people and 2,400 people engaged in the primary industry. The Village occupies a land area of 25.74 square kilometers, with an altitude of 1,200 meters, an annual average temperature of 17.6℃ and an annual precipitation of 1,329.6 mm. Besides, it is dominated by Lahu and Dai people, to be specific, including 2,259 Lahu, 1,045 Dai and 705 other nationalities.

Hekai Mountain is one of the six new tea mountains outside the Yangtze River with large and complete preservation area of ancient tea in Yunnan. The ancient tea trees in Hekai are mainly distributed in various villages and planted in patches and forests, mostly on slopes and mixed with rain forests, without practices of being pruned or dwarfed. Due to inconvenient transportation in previous years, the ecological environment is under this circumstance getting better and better. It is the main representative of the emerging ancient tea mountains after 2008. Because of the rich resources of ancient tea in the area, the locals have witnessed many newly-built large-scale primary

which is said to be the successor of the tea king. Later, a new tea king tree was named in the ancient tea garden of Banpo Village.

Totally, the area of Nannuoshan Tea Garden is more than 21,600 mu, including 12,000 mu of ancient tea gardens. Ancient tea trees are mainly distributed in 9 natural villages, and what most concentrating are 2,900 mu of tea gardens and 1,200 mu of ancient tea gardens in Zhulin Village. On top of that, Banpo Village holds 4,200 mu of tea gardens and 3,700 mu of ancient tea gardens and the Guniang Village is home to 3,500 mu of tea gardens and 1,500 mu of ancient tea gardens. Due to the wide distribution of Nannuoshan Ancient Tea Garden, the tastes often serve people various experience in the mouth.

Nannuoshan tea is one of the typical representatives of Menghai species of Yunnan big-leaf tea, the place is known as one of the six fresh-tea mountains outside the Yangtze River. Ancient tea in villages, either planted on slopes, or mixed with rain forests, benefits from the moderate ecological environment. There are many tea varieties in the tea area, and multiple excellent varieties were originated here, such as the well-known Nannuo Baihao, Yunkang tea series, Zijuan and other teas. Nannuoshan ancient trees have a long, compact and knotty symmetry and are relatively fine in all traits. When we take a sip of the tea, we can easily spot the golden and bright in color and feel a sense of bitterness immediately. However, the sweet taste returns faster than we could expect and the duration of astringency is longer than that of bitter taste, which serves as an amazing approach to quench our thirsty. The aroma of the tea liquor does not reach to an exaggerating point, despite it has lovely charm and good soaking resistance; the infused leaves are equipped with moisty yellow in color and nice flexibility.

1.11 Ancient Tea Forest of Jingmai Mountain

On September 17, 2023, the Cultural Landscape of Old Tea Forests of the Jingmai Mountain in *Pu'er* got inscribed onto the UNESCO World Heritage list, becoming China's 57th entry and Yunnan's 6th entry on that list.

Jingmai

Jingmai Mountain Ancient Tea Forest is located in Lancang Lahu Autonomous County, *Pu'er* City, with a total area of 19,095.74 hectares, of which 7,167.89 hectares are in the heritage area, involving Jingmai and Mangjing villages in Huimin Township, Lancang County. The heritage sites are inhabited by Blang, Dai, Hani, Wa, Han and other ethnic groups. For thousands of years the people of Blang, Dai, Hani and other ethnic groups in Jingmai Mountain have lived together in harmony, prospered and developed together with tea for generations, forming a colorful and unique tea culture with ethnic characteristics and creating a unique tea cultural landscape in Jingmai Mountain.

Jingmai Mountain is famous for its "Thousands of mu of Old Tea Forests", which is the largest, best-preserved and oldest cultivated tea forest in the world, and has the reputation of the "Natural Museum of World Tea History and Culture". Jingmai Mountain has over 3.2 million ancient tea trees and covers 28,000 mu, which is the physical example and typical representative

the main ethnic groups. Mengsong is a place name in Dai language, which means flat dam on a high mountain. The mountainous area occupies more than 95% in the total 492.67 square kilometers with Menghai County governing the township. It is situated at the southern edge of the Hengduan Mountains where terrain is inclined from northwest to southeast. Most of the mountains in the territory are in north-south slide. The highest point is in the west, with an altitude of 2,429 meters (also the highest point in the Prefecture), and the lowest point is at the intersection of Huiling River and Liusha River in the southeast, with an altitude of 772 meters and a relative altitude difference of 1,657 meters. Climatically speaking, the temperature is lower in areas 1,500-2,000 meters above sea level, with an annual average temperature of 16-17℃ and an annual rainfall of about 1,500 millimeters. Mengsong Ancient Tea Mountain, facing Nannuo Mountain across Liusha River, is one of the oldest ancient tea areas in Menghai. According to the analysis of dozens of oversized ancient tea trees left in Baotang Village of Mengsong, the history of tea planting of the ethnic minorities in Mengsong mountain area is as long as that of ethnic minorities in Nannuo mountain area.

There are more than 3,000 mu of scattered ancient tea gardens preserved in Mengsong ancient tea mountain, in which Baotang, about 10 kilometers away from the township government, is the most representative ancient tea village in Mengsong township. Most of Mengsong ancient tea gardens and villages near the gardens are owned by Lahu people. During the reign of Emperor Guangxu of the Qing Dynasty, Han people settled in Baotang and Nanben of Mengsong and traded tea as a means of living. Mensong tea is remained as symmetrical, slender and beautiful, with the buds distinct to tell.

Taking a sip of the tea, the aroma is lingeringly dulcet, the color is yellow and bright, the taste is slightly bitter and overtly astringent, but it melts in our mouth quite quickly. Besides, the general taste is full, rich, heavy and lasting as well, despite the fact that the tea liquor shadows them a bit. The infusing times are fewer than those of other tea in mountain peaks in Menghai and the infused leaves are yellow and flexible, with good freshness.

1.10 Tea Tree Planting in Nannuoshan Mountain

Nannuo Mountain is located next to the highway from Jinghong to Menghai, 24 kilometers away from Menghai County. It is a famous tea manufacturing area in Xishuangbanna, with the average altitude of 1,400 meters, the annual precipitation 1,500-1,750 millimeters, and the annual average temperature 16-18℃, which provides very suitable archetype for the growth of tea trees. The Nannuoshan Village Committee governs 30 natural villages, all of whose residents are Hani

Nannuoshan

people, The tea was majorly produced by a king of cultivated ancient tea trees known as "the King of Tea Trees", whose base diameter is 1.38 meters and the history of more than 800 years has witnessed the cycling pattern of its development. Unfortunately, it finished its mission in 1994. Besides, there was still a large tea tree with a stem diameter of more than 20 centimeters,

and insect pests, and without the need for human intervention.

The sun-dried green Maocha, one of the kinds of *Pu'er* tea, is made of fresh tea leaves picked from Pingzhai Ancient Tea Mountain. Its leaves are plump and strong with a bright blackish green color. And the dried tea has a lasting flavor with long leaves. The liquor of new-made tea has a bright green and yellow color. When the tea is boiled, mixed aroma of flowers comes first, then the fruity smell assails the nostril. Moreover, the liquor of sun-dried tea has a slight bitter and astringent taste when entering the mouth. After tasted carefully, it has mellow and thick, sweet and smooth taste, leaving a lasting and strong smell in the drinker's throat.

1.8 Malipo Ancient Tea Mountain

The growth environment of ancient tea trees in Malipo is located in the core area of three major biodiversity centers in China, with a variety of vegetation symbiosis and is recognized by biologists as "treasure land in southern areas". It enjoys a temperate and moist climate, and is surrounded by clouds and fogs. Moreover, it is rich in fertile yellow soil, a kind of medium-coarse-grained crystalline limestone that is formed from weathered granite. All of these provide a favorable environment for the growth of tea trees.

Malipo

Ancient Tea Mountain in Malipo, with a total area of over 30,000 acres, has various kinds of tea trees, amounting to more than 400,000 plants. The main varieties include local Baimao tea, *Pu'er* tea, etc., which are all fine ones. The mountain is a sea of tea trees that are over 100 years old and round 100 centimeters. And there is rich biodiversity and good vegetation.

There is abundant rainy and foggy weather, effective sunlight, warm and humid conditions, and tea trees have a long growth period. The tea buds are plump, hairy, tender and smooth, sweet and refreshing, with a mellow taste, good aftertaste, good tenderness retention. New shoots have abundant inclusions and wide fitness. The products have excellent quality characteristics such as "high floral aroma, sweet liquor and sweet aftertaste to produce saliva".

1.9 Tea Tree Planting in Mengsong Village

Mengsong Township is located in the east of Menghai County, adjacent to dozens of frontier villages. The township government is 23 kilometers away from the county seat and 39 kilometers away from Jinghong, capital of Xishuangbanna Dai Autonomous Prefecture. There are 21,467 people residing in 4,781 households altogether where Lahu and Hani are

Mengsong

and an annual precipitation of 1,374 millimeters, it is both the oldest and the largest Bulang ethnic village in Bulang Mountain, Menghai County. According to the stone tablets in the ancient temple of the village, the time when the Village was built happened to be the starting year of the Dai calendar, which claims a vivacious history of 1,371 years. In the ancient tea garden, trees engraved with vicissitudes of life have witnessed the long history of tea planting of the ancestors of the Bulang people.

The tea trees, so typical representatives of Menghai species with large leaves, are distributed and cultivated around the village. The age of tea trees is 100-500 years. Scrubby tea trees have been another mentionable trait in the Old Man'e Tea Village for decades, and the tea plucked from small tea trees is also known as sweet tea. Bitterness is a major feature of Old Man'e tea. Spring tea has a heavy raw taste and is extremely cold in its initial nature. It tastes bitter like, coupled with plump and thick texture, symmetrical and clear appearance, and has a clear bitter and cold flavor. Moreover, a lingering aroma could be developed after long time soaking. It is prosperously resistant to soaking with bitterness and a slight of astringency, but the bitterness could be holding rearwards and be lasting for a long time. Compared with bitter tea, the aroma of Old Man'e sweet tea is more mellow and higher, but it still generates obvious sense of bitterness and astringency, which notwithstanding vanishes faster and stimulates saliva well and better.

1.7 Pingzhai Ancient Tea Mountain

Pingzhai Ancient Tea Mountain is located in Fadou Township, in the southeast of Xichou County, Wenshan Zhuang and Miao Autonomous Prefecture, Yunnan Province. The Tropic of Cancer cuts through its core area, with north latitude 23°16′~23°32′ and east longitude 103°23′~104°92′. Ancient Tea Mountain covers an area of 109.7 square kilometers. It is located in the transitional zone between South China Hills and Yunnan-Guizhou Plateau, and the annual fog weather in the area reaches 146-198 days. Special geographical condition——low latitude and high altitude provides adequate sunshine for tea trees' growth. The perennial fog reflects the strong light into the cold spectral diffused light required for tea growth. Ancient Tea Mountain is adjacent to the Xiaoqiaogou National Nature Reserve, with a forest coverage rate of 87%, an annual average rainfall of 1,200mm, and an average annual relative humidity of 86%. This area is rich in biodiversity. Most tea trees are planted in native vegetation or secondary

Pingzhai

forests. In the tea field, the trees grow luxuriantly, with a carpet of leaves, humus layer and moss, which provide the best fertilizers for the growth of tea trees. Tall trees provide moderate shade and light, forming the best to growth of tea trees. A complete ecosystem witnesses low plant diseases

Also, nutrient accumulation decomposition and utilization are prone to emerge above the surface in this context, especially when soil organic content is up to more than 4.5% and humus thickness grows over 5 cm of the whole. Apart from that, the vegetation ecosystem in the distribution area is celebrated for ancient tea trees, with Toona sinensis, Cinnamomum camphora, Ficus microcarpa, Rhus, and Cinnamomum and other many aerial plants, lush trees and botanical gardens, as you can name, which constitute a good ecological environment and expedite an ideal place for tea planting.

Yiwu tea is famous for its mild perfume, soft with hardness. The dry tea trips are plump and tight with long and moist leaves. Its liquor color is bright yellow, and nectar smell has lasting strong aroma. The tea liquor is sweet and smooth in the mouth, well balanced in all aspects, with rich layers of taste changes, low bitterness and astringency, weak irritation, sweet aftertaste, full of flavor and high degree of soaking resistance.

1.5 Large-Leaf Tea Tree Planting in Mengku

At the junction of Lincang City, the west of Shuangjiang County and Gengma County, there lies one of the South-North branches of Hengduan Mountain System——Bangmashan Mountain.The main peak is Mengku Daxue Mountain, with an altitude of more than 3,200 meters. The famous wild and ancient tea community in Mengku is located at an altitude of 2,200-2,750 meters. Mengku wild tea community is a tea community with the highest altitude and the largest density found both at home and abroad, and a distribution area of more than 12,000 mu. Mengku large-leaf tea with some respects are oversized, tagged along long oval in appearance, dark green in color, and compressed scent in pleasure, with its buds and leaves plump, yellowish green and fuzzy. Spring tea boosts with one bud and two leaves and the dry sample of tea contains about 1.7% amino acids, 33.8% tea polyphenols, 18.2% catechins and 4.1% caffeine,

Mengku

providing a suitable desideratum for making black tea, green tea and *Pu'er* tea. Mengku Daxueshan *Pu'er* tea has a quality of thick taste, dense aroma and rich elements.

1.6 Tea Tree Planting in Laoman-e Village

The natural village is subordinate to Mengku Village, Bulangshan Township, Menghai County, Yunnan Province, and is a huge part in the mountainous area. Located in the northeast of Bulangshan Township, it is 16 kilometers away from Bulangshan Township Government. With a land covering 68.4 square kilometers, an altitude of 1,650 meters, an annual average temperature of 18-21℃

Laoman-e

as long as 60-110 cm. According to the annals of Mianning County at the end of the Qing Dynasty and the beginning of the Republic of China, "there are 6,000 or 7,000 tea growers in the county. Bangdong Township is specially remarkable of tea from Manlu and Xigui, of which Manlu tea enjoys the most fame in all aspects ". The Manglu, once called Manlu, has tea trees which are 3-4 or 5-6 tall high and are in a natural growing condition. Several tea trees with only a dead stump left in the trunk, but new branches with thick hoes growing from the bottom. The base girth of the big tea tree is about 80-90 cm. Traditionally, it is natural for tea-plucking manufacture and an ideal environment for tea growth. After a hundred years of artificial and unintentional creation, the branches are winding upward and the shape of the tea garden is jagged and becomes strange gradually, in a shape like a sleeping dragon and birds spreading their wings. Currently, the both purposes of being ornamental and functional in the tea garden hold availability for people who feel like climbing artificial mountains and plucking tea leaves.

When it comes to the color of the liquor, Xigui tea is demonstrated with light and clear yellow, showing overall impurities in the tea; the sweet taste is more obvious, the aroma is sharper, saliva is generated on both cheeks and bottom of tongue and ineluctably we feel slightly astringent on our tongue, which nevertheless melts quickly after we take several more sips; when infused for a modest time, crystal sugar aroma is gradually popping out, the water quality is becoming viscous, bitterness is vanishing bit by bit; still, the color of the tea liquor is almost unchanged, with integrated lovely and lingering charm emerging in long aftertaste; after total soaking, the water fades, the sweetness decreases slightly, though the sweet taste returns well, the rock-sugar like aroma remains. the dark green willow-strip-shaped infused leaves are flexible and bright.

1.4 Tea Tree Planting in Yiwu Mountain

Scattered in the east of the six chief tea mountains, close to the border between China and Laos, with an area of about 750 square kilometers, Yiwu Tea Mountain is the largest one with regards to both geography and production. An ancient tea garden is linked to it covering an area of more than 14,000 acre, mainly concentrated in willy-nilly villages. Yiwu Township borders Jiangcheng in the north, Yao District and Mengban in the south, Xiangming in Menglun in the west and Laos in the east. Notably, there are massive differences

Yiwu

in terms of altitude, climate, and thus resulting in rickety ecological environments falling in separate regions, which, if any, are tantamount to warm inclined climate. Yiwu boosts with plenty of sunshine and ample rainfall throughout the year as well as abundant high mountains with thick fog surrounding them. The land is known as ideally fertile and rainy enough to give birth to verdant tea trees under the blessing with moist penchant. When we consider the soil conditions of tropical and subtropical monsoon formed by weathering on sinuous and precipitous rocks, the soil in various places portentously shows slight acid reaction with PH in a range between 4.5 and 6.5.

It is adjacent to Lincang in the east, Baka in the south, Gengma in the west and Lincang in the north, with five villagers' groups under its jurisdiction, At present, 273 rural households are making their livelihood in this region, with a rural population of 1,064 and 955 labor forces. The altitude is 1,400-2,500 meters, the annual average temperature is 18-20℃, and the annual precipitation is 1,800 millimeters. The main ethnic groups are Dai, Lahu and Bulang people.

Originally, Bingdao tea marks as the source with a long and magnificent history of planting of Mengku tea species. The earliest history of planting tea can be traced back to the Ming Dynasty. Mengku tea is a national treasured tea, known as the authentic big-leaf tea or Yinghao big-leaf tea. Ancient tea mixed with trees, multiple slopes, few forests, and poor vegetation are typical in the Bingdao

Bingdao

tea area. Tea in the two villages in the eastern half of the mountain is basically crooked, waxy and astringent, but locals have the ability to produce tea with smooth fluid that lasts longer and tastes a slightly thinner. However, tea in the three villages in the western half of the mountain marks a distinct style, where the ancient tea trees exert no obvious bitterness. Besides, the aromatic flavor here is catching, the liquor is reaching to its full brightness and the reproduction of juice is cycling on a regular basis. What draws most people's attention is its unique crystal sugar rhyme. Comparatively, tea in the southern part has denser aroma and heavier bitterness than those of other villages. Besides, the reproduction of juice is slightly behind the latency and no obvious astringency is doped out but the overall picture of the tea shows a decent status.

1.3 Tea Tree Planting in Xigui Village

Known as part of the mountainous area, Xigui Village is subordinate to Bangdong Administrative Village, Bangdong Township, Linxiang District, Lincang City, Yunnan Province. It is 12 kilometers away from the village committee and 16 kilometers away from the township government. The land area is 3.82 square kilometers, the altitude is 704-1,043.4 meters, the annual average temperature is 21℃, and the annual precipitation is 1,200 millimeters.

Xigui

Xigui ancient tea, categorized as Bangdong big-leaf tea, is mostly distributed in the forestry and mountainous areas. The larger basegirth of the anciernt tea trees reaches

Section Four Famous Tea Tree Planting Mountains and *Pu'er* Tea Brands in Yunnan Province

1. Famous Tea Tree Planting Mountains in Yunnan Province

Distributed mainly in the five dominating tea-producing areas in Yunnan, the famous tea mountains in Yunnan are chiefly Xishuangbanna Tea Area, *Pu'er* Tea Area, Lincang Tea Area, Baoshan Tea Area and Wenshan Tea Area, with unique characteristics attached to each one and another. A wild profusion of vegetation, luxuriantly blooming florescence and splendidly flowing streams are observed all year round, and butterflies dancing and fluttering, as if exposed to fairyland in general. In light of the disparate geographical locations, quality of water, the taste of tea as well as regional climate on each mountain varies apiece. Let's take a closer look at them.

1.1 Laobanzhang

Laobanzhang is located in Bulangshan Township, Menghai County, Xishuangbanna Prefecture, Yunnan Province, with an altitude of more than 1,600 meters, the highest altitude of 1,900 meters and an average altitude of 1,700 meters. It belongs to the subtropical plateau monsoon climate zone, with no severe cold in winter and no scorching heat in summer, and only dry, wet and rainy seasons in a year.It has abundant

Laobanzhang

rainfall and fertile land, which is conducive to the growth of tea trees and the accumulation of nutrients. Laobanzhang is a Hani village in the north of Bulangshan Township Government, with an ancient tea garden of 4,490 mu. Since ancient times, villagers in Laobanzhang have used traditional ancient methods to artificially maintain the tea trees on which they depend for their livelihood, picking fresh leaves by hand according to folk customs, frying and kneading them with local methods. Laobanzhang *Pu'er* tea has a strong, thick and mellow tea flavor. It has always been regarded as "the king of tea" in *Pu'er* tea. Dried tea has fat strip, rich flavor, high quality and heavy bitterness but changes rapidly to sweetness. Between orchid and nectar smell, it has heavy perfume, which is very charming. Lasting remains the bottom of the cup strong aroma. In addition, Laobanzhang is very resistant to soaking. After more than 20 times of infusing, the tea flavor is still strong. exercises of immersion.

1.2 Tea Tree Planting in Bingdao Village

Under the jurisdiction of Mengku Town, Shuangjiang Lahu Bulang Dai Autonomous County, Bingdao Village is seated in the north of Mengku Town, 25 kilometers away from the position of Mengku Town Government and 44 kilometers away from the county seat.

brightness and darkness of infused tea leaves and the degree of leaf spreading are important factors for evaluating the quality of tea. Good infused tea leaves should be bright, tender, thick and slightly curled.

Terms for the evaluation of infused tea leaves:

Brown red: brown in red.

Red brown: reddish in brown.

Green and yellow: yellow is the main color, green in the middle.

Yellow and green: green is the main color with yellow.

Miscellaneous color: tea inclusions of different leaf colors, different shapes or multiple stalks, etc..

Soft: soft buds and leaves.

Tender and even: tender and even tea leaves.

It is believed that after the above evaluation and introduction, together with the personal experience of tea evaluators, we can better appreciate the taste of each *Pu'er* tea brew, and at the same time truly feel the charm of *Pu'er* tea.

3.4 Tasting and Appreciating

Good taste is one of the main factors that constitute the quality of tea products. Tea taste and aroma are closely related. Evaluation of tea can smell a variety of aromas, such as flowers, ripe chestnut aroma, often in the evaluation of tea taste can also be felt. Generally speaking and aroma is good, tea taste is also good. Tea aroma, tea taste identification difficulties when you can complement each other. Review of tea taste suitable temperature at about 50℃, the main difference between its intensity, strength, fresh, cool, mellow, and so on.

Intense: thick refers to the leaching of rich inclusions, with a sticky feeling; light is the opposite, fewer inclusions, thin and tasteless.

Strong and weak: strong refers to the tea liquor sucked into the mouth feels irritating or astringent strong, taste enhanced after tea liquor. Weak is the opposite—the entrance stimulation is weak, and taste is bland after tea liquor.

Fresh and crisp: fresh like eating fresh fruit feeling; crisp means crisp and refreshing.

Mellow: mellow indicates that the tea flavor is still strong, and the aftertaste is also cool, but the stimulation is less strong.

Harmony: indicates that the tea tastes bland and general.

Taste review terminology:

Intense: strong in the mouth, with a strong and persistent stimulation and a sweet aftertaste.

Mellow and thick: smooth and sweet in the mouth, with a long taste.

Mellow and calm: mellow and calm, with a slightly sweet aftertaste. Weaker in stimulation than mellow and stronger than calm.

Peaceful: tea with normal taste and weak stimulation.

Bland: slightly tea-like in the mouth with no aftertaste.

Watery taste: insufficient sense of concentration in the tea liquor, as thin as water.

Sweetness: sweetness at the base of the tongue and throat after drinking the tea liquor, and a feeling of nourishment.

Fresh and crisp: fresh and crisp.

Green and astringent: tea with a light and grassy taste.

Bitter base: bitter taste in the mouth and more bitter after taste.

3.5 Observing infused tea leaves

After tasting tea liquor, the leftover tea leaves were poured and placed in a clean plate to observe their tenderness, evenness and color. The tenderness, uniformity, fragmentation,

Red-brown: brown with red.

And the liquor color classification of *Pu'er* raw tea and *Pu'er* sun-dried tea differs, mainly:

Yellow-green: predominantly green with a yellowish tinge to the green.

Green-yellow: predominantly yellow with green in yellow.

Tender yellow: golden yellow with a soft white tinge.

Light yellow: not rich in inclusions, yellow and light.

Dark yellow: darker yellow, lusterless.

Bright yellow: yellow and shiny.

Orange-yellow: yellow with a slight reddish tinge.

In short, a good ripe *Pu'er* tea liquor color is red and thick and bright, and a good raw *Pu'er* tea and sun-dried *Pu'er* tea is yellow and transparent.

3.3 Smelling the Aroma of Tea

In addition to identifying the aroma, the main evaluation of the aroma of the pure and different, high and low, and the length.

Pure and different: pure refers to tea aroma, regional aroma, additional aroma. Different refers to the tea aroma impure or tainted with foreign odors, such as smoke and coke, rancidity, and oil smell.

High and low: mainly judged in terms of thickness, freshness, clarity, purity, flatness and coarseness.

Length: the degree of persistence of the aroma. Long means that the aroma can be smelled from hot to cold sniffing; the opposite is short.

Terms for aroma:

Millipore aroma: the aroma of the tea product with the buds and hairs showing.

Clear aroma: aroma fresh and sharp.

Aroma: aroma elegant, like the aroma of flowers.

Floral and fruity aroma: like fresh flowers, ripe fruit aroma.

Caramel aroma: sufficient drying or high fire power resulting in aroma with sugar aroma.

Sweet and pure: the aroma is pure but not high, but with sweetness.

Fragrant: elegant aroma with long-lasting aroma.

Intense: rich and persistent aroma with strong stimulation.

Stalks: young stalks.

Golden pekoe: tender buds with golden fuzz.

Slightly Tippy: with a lot of tea hair.

Dark red: red with darkness

Brownish auburn: auburn with brown.

Reddish-brown: red with auburn.

Blackish auburn: auburn with black.

Brownish-black: brownish-brown in color with a luster.

Black and moist: black and dark, seemingly coated with a layer of oil and bright.

Ebony: black bloom.

3.2 Observing Tea liquor

The evaluation of tea liquor color mainly covers three aspects: chroma, brightness and turbidity. The evaluation of liquor color should be timely, because polyphenols dissolved in hot water are easily oxidized and discolored after contacting with air.

Chromaticity: the color of tea liquor. It is related to the variety of tea tree and the aging of fresh leaves; the processing process determines the different liquor colors of various types of tea.

Brightness: refers to the degree of lightness and darkness of the tea liquor. Where the brightness of the tea liquor is good, the quality is also good.

Turbidity: refers to the degree of transparency of tea liquor. The liquor color is transparent without impurities, clear and transparent; the liquor color is turbid, with more floating impurities and no bottom.

Next, let's take a look at the specific method of identifying the liquor color of ripe *Pu'er* tea and raw *Pu'er* tea.

First of all, let's take a look at the liquor color classification of ripe *Pu'er* tea.

Orange-red: red with a yellow tinge.

Deep red: red and dark, lacking bright luster.

Chestnut red: red with dark brown. Also applied to the leaf base color of ripe *Pu'er* tea.

Red and thick: the liquor color is red and deep and thick. Red color of tea liquor and rich in inclusions.

Brownish red: red with brown.

ones, with mellow, thick, smooth, thin, and sharp mouthfeels. The aftertastes can be sweet, throat-moistening, refreshing and enriching the saliva. The joy of increased salivation is also subtle: increased salivation on the cheeks is like a torrential waterfall, rough and rapid; between the teeth and cheeks, it's like a gentle stream, soft and long-lasting; on the surface of the tongue, it's like warm, sweet dew, delicate and detailed; refreshing and enriching the saliva on the tongue is like a gentle, smooth, and peaceful stream. The process of tasting tea is also a process of enjoyment, balancing quality, strength, and time to truly savor the authentic flavor.

3. The Assessment of *Pu'er* Tea

As previously mentioned, when introducing the elements of tasting *Pu'er* tea, there are many types of *Pu'er* tea. They are categorized by the method of production into raw and ripe *Pu'er* tea; by storage method into dry-stored and wet-stored *Pu'er* tea; and by form into cake-shaped tea (Qizi cake-shaped tea), bowl-shaped tea, brick-shaped tea, and loose-leaf tea. Next, we will introduce the evaluation of *Pu'er* tea by examining the dry tea, observing the color of the tea liquor, smelling the aroma, and tasting the flavor.

3.1 Observing the Dried Tea

The shape, completeness, brightness and neatness will be assessed by observing dried tea.

Shape of the dried tea: the size, length, thickness, weight, and evenness.

Completeness of the dried tea: the completes of tea leaves.

Brightness of the dried tea: the color of the dried tea leaves.

Neatness: no weeds, seeds or other miscellaneous junk.

In short, good quality *Pu'er* tea is tightly-pressed, brightly-colored, and vitality-oriented.

Terms for describing the dried tea:

Completeness: the complete shape with no damage or breakage.

Balance of tightness and looseness: tea leaves are appropriately pressed.

Smoothness: the smooth surface.

Tip: tender tea leaves with shoots.

Thickness: the weight and thickness of the dried tea.

Lightness: the opposite of thickness.

Strength and Looseness: tender tea leaves are strong and loose.

Buds: fuzz on the tea bud.

Stems: tender stalks.

Appreciating the full flavor of *Pu'er* tea liquor requires certain skills. Do not drink it like a fish, as this approach precludes fully experiencing the tea's flavors. The key is to sip slowly, allowing the tea to gently circulate in the mouth, and then swallow delicately. When tasting the tea liquor, it's important to keep the oral cavity as open as possible, with lips closed and teeth apart, to create more space in the mouth. At the same time, it relaxes the muscles inside the mouth, creating more space between the tongue and the upper jaw, enabling the tea liquor to fully reach the lower teeth and the underside of the tongue. During the act of swallowing, as the space in the mouth narrows, the tea liquor and air beneath the tongue are compressed, creating a bubbling sensation known as the "refreshing and enriching the saliva." Truly appreciating tea goes beyond the quality of the tea and water; it requires a harmonious mindset, to deeply connect with the essence of the tea and elevate the experience to a more spiritual level.

By mastering these tasting techniques, when tasting *Pu'er* tea again, we can discern the rich historical character and unadorned nature of aged *Pu'er* tea, showing its essence.

2.4 Infusing Times Test

Infusing times refers to that the tea liquor color and taste without significant changes after infused for many times. This infusing times varies according to the type of tea. In everyday life, people commonly experience that bagged black tea, green tea, and scented tea are generally discarded after one infusing. This is because these tea leaves are cut and rolled during processing, which fully breaks down the leaf cells, the formation of granular or fine shape making it easy for the active ingredients to be extracted quickly during infusing. Ordinary green tea can usually be infused three or four times. The infusing times of tea leaves is related to their tenderness, but more importantly, it depends on the integrity of the leaves after processing. The more finely processed the leaves, the easier it is to extract the tea essence; the coarser and more intact the leaves, the slower and better the release of internal substances. Oolong teas like Tie guan yin are known for retaining their aroma for up to seven times of infusion , because it just can be infused seven or eight times. However, *Pu'er* tea is perhaps the most infusible among all teas. The lasting infusing times of *Pu'er* tea is due to the rich internal substances contained within the tea. Having grown for hundreds or even thousands of years, the buds and leaves of *Pu'er* tea accumulate rich nutrients, requiring infusing many times to fully release their essence. This is why *Pu'er* tea is known for its lasting infusing times. Under normal infusing conditions, ordinary *Pu'er* tea can be infused about ten to twenty times, with each infusing time revealing subtle changes. When raw *Pu'er* tea is infused for the frist time, it reveals a surface-level bitterness, as the leaves are not yet fully expanded. It is only after infused five or six times that the true inner quality of the tea emerges. By infusing ten times, the depth of the tea's character is revealed. When infusing ripe *Pu'er* tea, there is a rough, shallow taste, with a strong flavor and sometimes a residual pile-fermentation taste. However, after infused five or six times, the tea liquor becomes clear and wine-red in color, with a pure and strong taste, free from extraneous flavors, subtly sweet, delicate, and smooth, seemingly flowing down the throat without swallowing. By infusing twenty times, the color of the liquor gradually lightens, and the sweet like rock sugar water. Some ancient tree teas can even be infused more than forty times.

In summary, the tastes of *Pu'er* tea encompass sweet, sour, bitter, astringent, and fresh

(4) Refreshing and Enriching the Saliva at the Bottom of the Tongue

The phenomenon known as the "refreshing and enriching the saliva at the bottom of the tongue" occurs when tea liquor touches the area under the tongue and near the lower teeth, it feels like bubbles are coming out and stimulates saliva production. This effect is particularly obvious when enjoying well-aged *Pu'er* tea with a long period of alcoholization. As the tea interacts with the mouth and reaches the underside of the tongue, it gently produces saliva, and there will be the feeling of tiny bubbles continuously emerging. This is because during the process of alcoholization, tea polyphenols undergo a series of complex chemical reactions like oxidation, hydrolysis, synthesis, and fragmentation, which can't stimulate the cheeks or tongue to produce saliva, but some of the newly synthesized substances play a role in refreshing and enriching the saliva at the bottom of the tongue. The process here is subtler and more prolonged, resulting in a more refined and smoother sensation, and the sensation of refreshing and enriching the saliva becomes more pliable and tranquil. When tasting aged *Pu'er* tea, the tea liquor is very soft. Upon contact with the mouth and reaching the underside of the tongue, the bottom of the tongue slowly produces saliva, as if tiny bubbles are continuously emerging. This phenomenon of producing saliva under the tongue is the true essence of the refreshing and enriching the saliva at the bottom of the tongue.

melting. The embodiment of moistening is like the sweetness—it is the lubrication and moist feeling after tasting the tea liquor, and it can be said that moistening is the comprehensive reflection of the taster's taste of the tea liquor. The experience of moistening not only means that the tea liquor is full of taste, but also means that the taste is smooth and does not appear to be stuck or scraped in the throat. The taster will form the experience of melting only after adapting to the taste and texture of the tea liquor. Most of the *Pu'er* teas, after proper aging, can reach the realm of "throat moistening, breaking the beredom". The new tea, on the other hand, has higher requirements on the raw materials and processing technology.

Refreshing and enriching saliva: "Jin" refers to saliva, and "Shengjin" means the secretion of sativa in the mouth. The raw material of *Pu'er* tea is large-leaf sun-dried maocha, so the tea is rich in internal components, especially the high content of ester-type catechins (EGCG, ECG, etc.), from the astringency to the production of saliva, whose function is particularly strong. Some of the inferior tea products, after drunk always feel that the inside of the mouth is rolled up, the cheek muscles are spasmodic and uncomfortable, and the tongue moss is thickened, but there is no feeling of generating fluid. This kind of astringency but can not produce the Shengjin, called "astringency can not be melted". The specific subdivision of the production of Jin is the production of Jin on both cheeks, the production of Jin on the teeth and cheeks, the production of Jin on the surface of the tongue, and the production of Jin at the bottom of the tongue (under the tongue).

(1) Refreshing and enriching saliva on the cheeks

Among them, the most intense one is saliva refreshing and enriching. It refers to the secretion of saliva after the tea liquid is consumed because the astringent substance stimulates the lining of both sides of the mouth. This type of saliva, fulling the entire mouth, is large in quantity and strong in taste. Early spring tea or tender astringent tea refresh and enrich more saliva in cheeks. If the body loses too much water, it's better to choose tea with the effect of refreshing and enriching saliva on cheeks.Tea is easy to brew and has a better effect of quenching the thirst.

(2) Refreshing and enriching the saliva on the teeth

During the process of drinking *Pu'er* tea, the tea liquid flows in the mouth, and the tannins in it stimulate the lining between the cheeks and the teeth to secrete saliva. Different positions generating saliva lead to different feelings. The saliva produced at the cheeks is like a waterfall, rough and sharp; while the saliva produced at the teeth is like a trickle of streams. The place where it is moistened is warm, sweet and smooth, which is more obvious in the slightly fermented *Pu'er* ripe tea.

(3) Refreshing and enriching saliva on the tongue

Tea products whose astringency disappears faster will produce moist liquid on the tongue after drunk, which is called refreshing and enriching saliva on the tongue. Specifically, after the tea liquid is swallowed, saliva is slowly secreted in the mouth, and the tongue is very warm and delicate. At the same time, the surface of the tongue seems to be secreting saliva continuously, which then flows to the oral cavity on both sides of the tongue. Usually, good-quality *Pu'er* tea can make the tongue produce saliva.

Thickness and thinness: Thickness means that the tea liquor is rich in inclusions, high in water leachate content, resistant to steeping, thick and heavy, with a sense of texture, as the saying goes, "it can hold your tongue". Spring tea is thick while autumn tea is thin. If you put more tea leaves in the tea set when brewing and cover the cup, tea liquor will be stronger in flavor and thicker in taste. On the contrary, if you put fewer tea leaves, the liquor will be blander. Therefore, to compare the thickness and thinness of the tea liquor, it is necessary to obtain the tea liquor under the same brewing technique and taste it.

Sharpness is called throat scraping. The main reason is that the contents of tea liquor are not balanced. Some of the inclusions in the tea liquor are too many and some are too few, not perfectly balancing the various complex flavours, making one or more of the offbeat, strong flavours over-stimulating the taste and touch senses, making the taster feel like a sharp blade in the throat, collecting and scraping our throats. The sharpness can also be caused by the quality of the water when the salt content in the water is relatively high. When the tea liquor is relatively tasteless, the irritation of these salt ions on the throat will appear, causing the taster to feel the sharpness of the tea. Sharpness can also be influenced by processing methods: Over-rubbed or physically damaged tea leaves have a higher breakage rate than liquor but are not resistant to infusing. The first few infusions are thick but will have a scratchy feel as the inclusions leach out in uncoordinated proportions and the taste becomes very thin and monotonous after a few infusions.

(3) The sense of aftertaste

The taste and texture of *Pu'er* tea are the real feelings and experiences of tea drinkers, but the taste is not only these, but also the aftertaste.

It mainly includes three parts: returning sweetness, moistening throat and refreshing and enriching saliva. These reactions are the gifts given by tea to the taster, as well as the sublimation of feeling after drinking and the enjoyment of the soul.

Returning sweetness: Returning sweetness is different from sweetness. Sweetness is formed when the tea liquor soaks the tongue, while returning sweetness is the sweetness of the taster's own feeling after tasting the tea liquor. The experience of returning sweetness is more introverted, delicate and long. The sweetness that appears in the mouth after tasting the tea is the expression of returning sweetness. Some poor-quality *Pu'er* teas have bitter taste but no returning sweetness, while some high-quality *Pu'er* teas have returning sweetness without bitterness.

Returning bitterness: Returning bitterness is the opposite of returning sweetness. After drinking, the bitterness remains in the throat for a long time. There are two kinds of bitterness in *Pu'er* tea tasting: one is bitter in the mouth, and the bitterness turns into sweetness, also known as bitter first and then sweet; the other is the tea liquor is not bitter in the mouth, and then it turns bitter and does not dissipate for a long time. Most of the tea with bitterness is inferior tea, wet-storage tea, rough old tea essence with insufficient greening, etc. In the tasting process, the tea is not normal and should not be collected because of improper processing and improper preservation.

Moistening: moisteningt is the sublimation of smoothness, and there is a sinoothness

acids and theaflavins, as well as trace soluble peptides, nucleotides, succinic acid and other components.

Acidity: Acidity is a bad taste of *Pu'er* Tea. Low-quality *Pu'er* tea tastes sour. It can make teeth and cheeks tense and close. It is due to improper drying of raw materials, inadequate heaping temperature or improper storage during fermentation.

(2) The sense of mouth (mouthfeel)

Mouthfeel is based on taste, which integrates other sensory nerve communities in the oral cavity and makes comprehensive sensory evaluation. The formation of taste is not only influenced by the chemical composition of tea liquor, but also by a series of physical factors such as density and stickiness, temperature.

Smoothness: Sensorially, it is like a fluid gently brushing the surface of your tongue and entering your throat, a soothing and restful sensation. The main substances affecting the smoothness of *Pu'er* tea are soluble sugars, oligosaccharides and pectins. The sugars can wrap around the polyphenols, reducing the "astringency" and increasing the "smoothness" of the tea liquor. In the tasting of *Pu'er* tea, smoothness is an important indicator of the quality of *Pu'er* tea. A tea liquor that is not smooth is either "stuck in the throat" or bland, giving people an uncomfortable feeling.

Generally, in old tea or mature tea, smoothness is more obvious, while in sun-dried tea, especially the new, smoothness is very weak

Melting: This refers to the speed at which the taste sensation of the tea liquor changes as it flows through the mouth. The description of "melting" is rather abstract, and what tea tasters generally mean by melting and not melting is how long the tea liquor stays in the mouth after it has been tasted. "In the mouth it melts" means that the taste of the tea liquor can naturally dissipate after a few seconds in the mouth, and the aftertaste is endless; "in the mouth it is difficult to melt" is the taste of the tea liquor takes a long time to stay on the tongue and is difficult to dissipate, dominating our palate, affecting the true taste of the tea liquor. The "hard to melt" tea liquor is one that stays on the tongue for a long time, dominating our palate and affecting the true taste of the tea. Ripened tea is easier to melt than sun-dried tea, and after the old tea has been mellowed over time, the quality is mellow and melts faster.

Alive: Compared to "smoothness", the taste of live tea is more refreshing and dynamic to the taster. *Pu'er* tea that has been refined over time undergoes a complex transformation of its internal components to produce a strong active taste when brewed into tea liquor. Dry-aged *Pu'er* teas that have been mellowed have undergone various chemical reactions such as hydrolysis, cleavage and oxidation, resulting in an increase in water-soluble substances and a decrease in molecular weight, and the rapid transformation of substances in the tea liquor, which also allows for the best expression of active taste.

Watery: Coarse raw materials, less flavoring substances or high humidity in the tea during fermentation lead to the feeling of "water" in tea liquor. After other tastes melt away, there is a sense of reactively tasteless, like drinking a cup of water. The taste of water can identify the age of the raw material, the fermentation process, and the proper preservation.

(1) The sense of tasting

Sweetness: It is a flavor that makes people feel sweet but not greasy. Sweetness is caused by the hydrolysis or cracking of carbohydrates to form sugars or oligosaccharides.

Sweetness is not only enjoyed by children, but also by adults who salivate over sugar. However, the sweetness of strong sugar is often loved and feared, whereas the light sweetness in tea is so elegant and harmless to health. The light sweetness elevates *Pu'er* tea tasting to the realm of art. *Pu'er* tea is a large-leaf tea, which is relatively saturated and thick in composition. After long-term aging, the bitter and astringent flavours slowly diminish due to oxidation, or even disappear completely, while the sugar remains in the tea leaves and is slowly released into the *Pu'er* tea after brewing, giving it a sweet taste. In good *Pu'er* tea, the sweetness becomes stronger and stronger as it is brewed. The sweetness in *Pu'er* tea liquor is pure and elegant, and best represents the true nature of *Pu'er* tea.

Bitterness: Bitterness is a certain taste of *Pu'er* tea, and it is one of the criteria for identifying the quality of tea. Only after bitterness can there be aftertaste. And it brings the *Pu'er* tea taster that revelation of the true path. The reason why *Pu'er* tea is bitter is because it contains caffeine. The reason why tea is so refreshing is that this caffeine has a stimulating effect on the human nervous system. A truly healthy *Pu'er* tea tasting does not seek to refresh the mind through bitterness, but rather to achieve a sweetness and throat effect from the slightly bitter tea liquor. *Pu'er* tea made from young and tender-grade tea leaves has a bitter taste. The handling of bitterness is controlled by the brewing method, and the appropriate bitterness of the tea liquor is also determined by the taster's acceptance of the bitterness.

A good *Pu'er* tea has a very distinct "bitter then sweet" sensation. The sweetness of the aftertaste is one of the characteristics of *Pu'er* tea and one of the reasons why people like to drink it. The sweetness and longevity of the aftertaste is one of the factors to identify a tea. For the famous Lao Banzhang and Jingmai ancient teas, the sweet and smooth sensation in the mouth and throat can last for one or two hours after drinking the tea if there is nothing else to disturb the taste sensation.

Astringency: Astringency is the original taste of tea. It is caused by the complexation of lipid ester type catechins with proteins in the cells of the mouth, which makes the tongue feel thicker, the lining of the mouth thicker and something sticks to it. It is often heard that not being bitter or astringent is not tea. In fact, the bitterness gradually fades away in aged *Pu'er* teas that have been aged for more than six or seven decades. There are yang *Pu'er* teas with a stronger taste and yin flexible *Pu'er* with a more docile taste. Which are the rigid ones? Which are flexible? It is by its degree of bitterness that the most specific method of identification is. The astringency of tea is due to the presence of tea tannins. *Pu'er* tea is made from large-leaf tea essence, which contains more tea tannins than ordinary tea leaves, so the taste of new sun-dried tea is very strong, and the astringent taste is particularly strong. Adequate astringency is acceptable to tea tasters as it will cause the muscles in the mouth to tighten up and produce a tasting effect. Astringency increases the stiffness of the *Pu'er* tea liquor and also satisfies the tasters the heavier taste. Brewing both bitterness and astringency requires attention to its skill and personal acceptability.

Freshness: Fresh includes fresh, slightly sweet, elegant, refreshing. It is caused by amino

(5) The flavor of assorted aromas

Presenting the aroma. With the increase of storage years, the aroma changes from fresh aroma→fruit aroma→flower aroma→honey aroma→wood aroma→old aroma, the volatility of the aroma changes from "Yang" to "stable", and the molecular structure of the chemical components also gradually increases, which is caused by the oxidation and polymerization of compounds in the aging process. Ancient tree tea usually presents floral and honey-scents, and at the bottom of the cup it also leaves an obvious honey-sweet smell, and this kind of tea converts quickly during storage.

(6) Other tastes

A direct cause of the production of saliva is the stimulation of the oral cavity by various compounds in the tea liquid and the excitation of the salivary secretion center. Often, saliva continues to be secreted even after the stimulation stops, more due to the excitation of theanine to the parasympathetic nerves, which leads to more sustained gastrointestinal peristalsis and salivation. The continuous secretion of saliva moistens the mouth and throat, glycoproteins secreted by mucilaginous follicles in the salivary glands work with odor-presenting substances in the liquid to produce such feelings as aftertaste and throat moistening. Saliva can not only keep the mouth clean, but also help digestion, and protect the gastric mucous membrane. Ancient people attached great importance to the relationship between saliva and health, and the saliva produced after tasting tea liquid is regarded as the best drink.

Pu'er tea usually has several tastes such as sweet, bitter, astringent, sour and fresh, as well as smooth, cool, thick, thin and sharp tastes. At the same time, *Pu'er* tea also makes people have a sense of sweetness, and throat moisturization. The taste of *Pu'er* tea is formed by the combination of such taste, mouthfeel, and afteraste. All kinds of feelings may exist alone in a certain bubble of *Pu'er* tea, or may coexist, so it needs to be savored carefully in the process of tasting.

tea, the tea liquor will feel uncomfortable in the mouth, producing a dry and parched feeling in the throat, and even affecting swallowing if it is strong.

Taste, is a comprehensive subjective feeling formed by various stimuli produced by the senses of taste, smell and touch on the tea leaves and tea liquor. The taste of *Pu'er* tea originates from its water leachate, which is the basis of the essence of the tea leaves. Usually, the water leachate of *Pu'er* tea is 30%-50%, and the taste of different categories of substances has their own characteristics.

(1) The flavor of tea polyphenols

The taste in tea liquor is tannic acid, which is characterized by astringency. Astringency is a sense produced by the coagulation of tannins in the mouth. The chemical composition of tannic acid is complex and varies greatly depending on the raw materials. It can be divided into two major categories: hydrolyzable tannins (also known as ester catechins) and condensed tannins. The former is more irritating, the smell is obvious, and the mouth feels "rough"; the latter is weakly irritating, making the mouth feel "cool".

A tea of good quality stings the tongue at the entrance, but soon releases, a sensation known as "melting", which does not leave an overly astringent taste even after the temperature of the tea liquor has decreased. Some tea people use the strength of the "graps" on the tongue and the length of time it takes to "melt" as one of the criteria for judging the quality of tea leaves.

(2) The flavor of alkaloids

The taste is bitter. The bitterness is the basis of the "sweetness" and the production of the thirsty is the source of the "sweetness". This is due to the "illusion" of the taste sensation, which is exactly what people want. The degree of bitterness in the mouth and how quickly or slowly it dissipates is also a factor in judging the quality of the tea. If the bitterness does not dissipate or is too strong, it can be off-putting.

(3) The flavor of amino acids

They have a variety of performances and they have a strong synergy with other taste substances. The fresh, sour and sweet taste of amino acids will make people thirsty. Freshness: theanine, glutamic acid, aspartic acid; sweetness: glycine, alanine, etc.; sourness: glutamic acid, aspartic acid, etc; aroma (floral): glutamic acid, alanine.

(4) The flavor of sweetness

The taste is sweet and the smell is sweet. Sweetness has a great influence on the taste. In human instinctive needs, sugar is the most important. The taste and smell are very sensitive to sweetness. Sweetness can make people feel happy. Pectin in sugar plays an important role in mouthfeel. Pectin has the highest content in tea with moderate tenderness, 3% to 5% of dry tea. During the aging process, pectin can be degraded into water-soluble carbohydrates to increase the taste.

The aroma characteristics of *Pu'er* tea can be accurately identified by combining the three stages of hot smelling, warm smelling and cold smelling (Table 4-3).

Appreciating the aroma of *Pu'er* tea is a kind of spiritual enjoyment. In order to better smell aroma, it is advisable to choose larger procelain cups such as Justice Mug (Gongdao Mug). The inner wall of porcelain utensils is easier to hang aroma than glass utensils, and the inner volume of cups is larger, which can gather more tea aroma and make tea aroma.

Good *Pu'er* tea has pure and delicate aroma, elegant and coordinated, which can make people relaxed and happy. The leftover tea leaves have distinct and lasting aroma. Chinese people believe that we drink tea with heart and soul. Tea is considered as a living thing which requires humankinds to respect and appreciate it. When we brew, infuse, taste and smell *Pu'er* tea, we are actually communicating with nature seeking harmony between man and nature.

It should be mentioned that there is a major distinction between mildew odor and mellow aroma smelled and tasted from *Pu'er* tea. Mildew odor is an unpleasant scent which spoils taste. Mellow aroma is the result of fermented *Pu'er* tea, which is pleasant and enjoyable. Only high quality *Pu'er* tea can produce mellow aroma.

Table 4-3 Methods and Techniques for Identifying the Aroma of *Pu'er* Tea

Methods and Techniques for Identifying	Key Points	Precautions
Smell at a hot temperature	Aroma purity, type and level.	When the leaf temperature is above 65℃, it is easiest to identify whether the tea leaves have peculiar smell.
Smell at a warm temperature	Mainly to distinguish the type and quality of aroma.	When the temperature of the bottom of the leaf is about 55℃, it is easiest to identify the type of aroma.
Smell at a cold temperature	Mainly to distinguish the lasting degree of tea aroma.	When the leaf temperature is below 30℃, the aftertaste of tea aroma can be distinguished, and the higher is the better.

2.3 Taste

As the old saying goes, "aftertaste is called rhyme". Aftertaste refers to the balanced proportion of various taste substances in tea liquor, refreshing and comfortable entrance, thick and fragrant taste with level changes, which makes people have a certain feeling beyond taste pleasantly. This is the deeper enjoyment that *Pu'er* tea can bring to people. This feeling may make people have some beautiful artistic conception in the process of drinking tea, and this artistic conception can not only purify the mind, but also make people detached.

The glycation of aged *Pu'er* tea during the aging process causes the monosaccharides transformed by the tea body to oxidise and polymerise into polysaccharides, making its liquor sweet in the mouth for a long time, thus moistening the throat and relieving the feeling of thirst. After swallowing the tea liquor through the mouth, the saliva in the mouth will be secreted slowly and the tongue will feel very moist. On the contrary, with poor quality *Pu'er*

but it is more elegant and lighter. This aroma of *Pu'er* tea mainly comes from the aging of the tea stalks, and with more stalks it is more prominent in tea leaves in the late aging stage. Tea products such as Hongyin *Pu'er* tea (It refers to the Chinese character of "cha" which is encircled by eight "zhong" on the package are red) and Luyin *Pu'er* tea (It is similar to Hongyin *Pu'er*, but "cha" on the package is green) has a prominent woody aroma.

Mellow aroma: Mellow aroma is the real and heart-felt enjoyment of *Pu'er* Tea. Every tea lover knows that mellow aroma is the most precious element in *Pu'er* Tea. The older *Pu'er* tea becomes, the more mellow fragrant it possesses. Mellow aroma demonstrates the breath of time and history. It feels like a living old wine. It is light and elegant, low and lingering, with rosemary wrapped around, which makes tea drinkers happy, peaceful and intoxicated. Mellow aroma is the aroma of *Pu'er* Tea which gradually changes with the passage of time. People feel addicted when they smell mellow aroma. Mellow aroma is remarkable in the time-honored *Pu'er* tea, with a long, light and lingering charm, which seems to be fanscinaing.

Aroma smelling is generally achieved by three approaches: hot smelling, warm smelling and cold smelling.

(1) Hot smelling: Smell when tea liquor is hot

Hot smell refers to the instant smell of tea liquor brewed. At this time, it is the most easy to distinguish whether there is any odor, such as stale odor, mildew odor or other odors. As the temperature decreases, the odor emits, and the sensibility of the sense of smell to the odor also decreases. Therefore, when smelling hot, we should mainly distinguish whether the aroma is pure or not.

Because the aromatic substances contained in the infused tea can be fully volatilized under the action of heat, some bad odors can also be volatilized. Therefore, it is the easiest to distinguish the aroma purity of tea by smelling the leftover tea leaves while it is hot. One hand holds the cup with tea liquor, the other hand uncovers the cup lid, and sniffs deeply near the cup with the nose. In order to correctly distinguish the type, the level and the duration of tea aroma, it should be repeated once or twice, but the time of each smell should not be too long, because people's sense of smell is easy to be fatigue. When smelling aroma, shake the leftover tea leaves several times.

(2) Warm smelling: Smell when tea liquor is lukewarm

Warm smelling refers to smelling the aroma after hot smelling and seeing the liquor color. At this time, the temperature of the tea cup decreases and the hand feels slightly warm. When smelling warmly, the aroma is not hot or cool. It is easy to distinguish the type, concentration and intensity of the aroma.

(3) Cold smelling: Smell after sipping tea liquor

Cold smelling refers to smelling after sipping tea liquor. At this time, the temperature of tea cup has dropped to room temperature, and the feeling of the hand has cooled. We need deeply smell and carefully distinguish whether there is still a residual aroma. If there is still aftertaste at this time, it is the performance of good quality, that is, the permanence of the aroma is good.

Honey aroma: There are three kinds of honey aroma in *Pu'er* tea: flower honey aroma (nectar aroma), fruit honey aroma and bee honey aroma. Among them, the nectar is like pollen honey, sweet and stimulating, which reveals a burst of sweetness. Tea products with nectar aroma are gradually exposed after a certain degree of aging. Sweet and elegant fruit honey is the typical original aroma of *Pu'er* tea. Bee honey aroma is more remarkable in ripe *Pu'er* tea, mainly due to the fermentation of tea.

Flower and fruit aroma: The common flower and fruit aromas in *Pu'er* tea include the aroma of rose, rice, orchids, osmanthus, plum, etc. There are also many unknown and prominent wildflower aromas. There is a wide range of floral and fruity aroma types in *Pu'er* tea, and due to different regional environments and initial manufacturing processes, the types of floral aroma are also different. Many teas have typical regional aroma, for instance, Laoman'e tea in Banzhang Village has rice aroma; tea in Bulang Mountain and Bada Mountain presents a typical plum aroma; tea in Nannuo Mountain presents elegant glutinous rice aroma; Dabai tea and Gelang tea in Jinggu County present rose aroma, and tea in Fengqing County has a typical orchid-type black tea aroma, and so on.

Sweet aroma can be divided into sugar aroma and pure sweet aroma. The former is like the aroma of caramel and brown sugar, while the later is sugar-free aroma. Sweet aroma is the aroma of *Pu'er* ripe tea, which is presented by tea polysaccharides, oligosaccharides and monosaccharides formed after the degradation of a large amount of cellulose during the fermentation process. If tasting carefully, you will find that different teas have different sweet aroma, which can be used as a method to distinguish the tea.

Woody aroma: The woody aroma in tea, mainly due to the degradation of lignin, comes from sesquiterpenes such as nerolidol and 4-vinyl phenol. It smells like the aroma of wood,

Camphor aroma: There are many forests of tall camphor trees all over Yunnan. Most of these camphor trees are several meters high. The space under the camphor tree is most suitable for tea planting and growth. Camphor tree can provide shade for tea trees. Tea trees can help preventing the occurrence of pests and diseases under the camphor tree environment. For example, there are many spiders on camphor branches and leaves, which will drop silk and eat small green leaf insects in tea trees. Tea roots and camphor roots grow alternately in the underground, camphor branches and leaves also emit camphor, tea trees directly absorb camphor stored in the leaves, so *Pu'er* tea has a unique camphor aroma. However, it is actually due to the influence of ecological environment such as light and soil in the growth of tea tree that it leads to the corresponding aroma precursor substances produced in the growth and metabolism process of tea tree. After the influence of specific processing technology, *Pu'er* tea finally produces unique camphor aroma.

Lotus aroma: Tippy tea is the one picked before Pure Brightness in April (one of the 24 Solar Terms of the Chinese Lunar Calendar). Its new color is tender, green and lovely. After fermentation, young bud tea leaves give of strong green leaf aroma, naturally leaving a light lotus aroma. Before brewing, one can smell the light lotus aroma from loose *Pu'er* tea. The infusing time can directly affect the lotus aroma of *Pu'er* tea. It is advisable to brew it with fresh and good water. The soft water quality is the best. The water temperature should be boiling hot when infusing, and it is more appropriate to pour it quickly. When tea liquor is drunk in the mouth, it stays for a moment and opens the upper jaw in front of the throat. A lotus aroma enters the nasal cavity through the upper jaw. Under the sensation of smell, it emits light lotus aroma, which is elegant. It narrates the romantic charm of *Pu'er* tea and arouses the sensibility of beauty.

Orchid aroma: Orchid aroma combines the beauty of lotus and camphor, and is more implicit. Generally, dried tea without infusing is not easy to get the scent of orchid aroma. "Aroma in the nine- hundred mu orchid garden, round as the full moon in the mid-autumn" is the most beautiful sentence describing the orchid aroma of *Pu'er* tea. "Aroma in the nine-hundred mu orchid garden" is a metaphor of vast orchid garden, which can produce orchid aroma. "Round as the full moon in mid-autumn" refers to cake *Pu'er* tea is round like the round and beautiful moon in autumn. The meaning of this sentence implies that cake *Pu'er* tea is as fragrant as a vast orchid garden, and as plumpy as the full moon in mid-autumn.

Pure tea aroma: Pure tea aroma is the most common aroma in *Pu'er* tea. The aroma of tea products made of young and tender tea buds is the most obvious. It is not unique to flower, fruit, honey, etc., but is elegant and original. Pure flavor *Pu'er* tea mainly depends on its raw materials and production season. Generally speaking, the aroma of spring tea and young tea is more obvious, such as Dayi Spring Early, the aroma is very prominent.

Jujube aroma: Jujube aroma mainly includes green jujube aroma and red jujube aroma. Jujube aroma belongs to the flower and fruit aroma category, which is generally contained in some regional tea products of sun-dried green tea, and the distinction is more obvious. Jujube aroma for cooked tea contains sweet and red jujube smell. Generally speaking, jujube tea is caused by moderate fermentation. The newly fermented semi-finished product is light, but the fine product can still be smelled. After the product is aged in the later stage, the aroma of jujube tea will gradually become prominent.

color will be relatively thick and dark, close to the mildly fermented pile cooked tea. The color of infused tea leaves of *Pu'er* tea is mostly dark chestnut or black, and the texture of the leaves is thin and hard. However, some ripe tea will be very close to the sun-dried green tea if the pile time is short, the degree of fermentation is light.

In addition, when appreciating the tenderness of the infused tea leaves, we should prevent two kinds of illusions: one is we usually mistake some varieties of fat tea and long internode characteristics for coarse old strips; second, *Pu'er* tea stored in humid places has dark brown color and folded infused tea leaves, which will be considered as old tea compared with *Pu'er* tea stored in dry places tea with the same tenderness.

The infused tea leaves of sundried green tea

The infused tea leaves of aged raw *Pu'er* tea **The infused tea leaves of ripe *Pu'er* tea**

2.2. Aroma

Aroma is the soul of tea, and aroma is the eternal charm of *Pu'er* tea. There is a poem complimenting *Pu'er* tea, "The exotic tea made in South Yunnan can purify one's mind with its aroma". The aroma of *Pu'er* tea is quite rich. The aroma of raw *Pu'er* tea is elegant and remote, while the aroma of ripe *Pu'er* tea is stale, implicit and changeable. By infusing tea leaves with water, the aromatic substances contained in tea leaves can be volatilized, which stimulates senses of people. Because of the planting differences in the ecological environment, tree age, latitude, altitude, soil composition and storage time, the substances accumulated in tea leaves are diverse, resulting in the diverse aromas of *Pu'er* tea. The following summarizes the major types of aromas, which are camphor aroma, lotus aroma, orchid aroma, pure tea aroma, jujube aroma, honey aroma, assorted flower and fruit aroma, sweet aroma, wood aroma and mellow aroma.

It is also necessary to pay attention to identifying whether the color of *Pu'er* tea is normal. Normal color means that it has the proper color of the tea, for example, raw *Pu'er* tea is yellow green, dark green, or green. The appearance of ripe *Pu'er* tea is brown, red, black, black brown or dark brown, etc. If the color of raw tea is black brown or dark brown, the quality must be bad. Similarly, if the color of ripe tea is dark green or cyanine, the quality is not good. This is the primary condition for evaluating the color of *Pu'er* tea, followed by the fresh, dull and mixed color. When evaluating the color of raw and ripe tea, attention should be paid to the freshness of the color, that is, the color is smooth and energetic. At the same time, it should see whether the overall shape of tea is uniform, the color is harmonious, and whether there are other colors mixed together. Tea color observation also includes looking at whether there are fungus stalks and leaves. As time goes by, the color will gradually change to brown green, yellow brown, dark brown. The color of ripe *Pu'er* is better with brown and red and even glossy while the tea with dark color or mildew are bad.

(2)Tea liquor color observation

We are going to observe the color of tea liquor, which is one of the manifestations of production technology, storage time and storage conditions of tea. The color of ripe *Pu'er* tea liquor is red, orange-yellow, bright and transparent. Unclear and cloudy liquor indicates the bad quality of ripe *Pu'er* tea. When raw *Pu'er* tea is infused for more than ten seconds, the color of the tea liquor is yellow-green. After 1 minute, the color of the tea liquor would turn golden. Raw *Pu'er* tea stored for a certain year will change its liquor color from green to yellow-green in about 5 years, from yellow-green to golden in 5 to 10 years, and from golden to yellow-red in 10 years.

When appreciating the liquor color of *Pu'er* tea, it is better to choose glass teaware. After pouring 1/3 cup of tea liquor into a glass cup, we can raise the cup along eyebrows, and incline the cup mouth 45 degrees inward, so as to better observe the color and lustre of the tea liquor. When appreciating the shade and brightness of the liquor, the position of the teacup should be changed frequently so as to avoid the impact of different light intensities.

(3) Infused tea leaves observation

We are going to observe the leftover tea leaves. Check their color, hardness, softness, purity, tenderness and lustre. Smell the aroma and observe the color of the infused tea leaves. For example, the infused tea leaves of 10 years, sun-dried green *Pu'er* tea look golden yellow within 10 years. If it's more than 10 years, the infused tea leaves tend to be golden chestnut. With time passing, the infused tea leaves of ripe *Pu'er* tea gradually turn to be black and brown. Leaves are soft, tender, flexible, and curly.

Regarding the infused tea leaves of *Pu'er* tea, it can be evaluated from the following two aspects: one is to distinguish the softness and elasticity of the infused tea leaves by touching; the other is to distinguish the tenderness, evenness, color and development of the infused tea leaves by eyes; at the same time, it can also observe whether there are other sundry things mixed in. Good infused tea leaves should be bright, tender and soft. The infused tea leaves of *Pu'er* tea are yellowish-green or yellow, and the leaves are soft and full of freshness. There is also some sun-dried green tea in the production process, such as tea after rolling, not immediately dry, delayed for a long time, the infused tea leaves will be dark brown, liquor leaves, that is, watch the tender, uniform and complete degree of tea after brewing, but also see whether there are flowers and impurities and whether there are burnt spots, red tendons, red stems and other phenomena. When buying *Pu'er* tea, we can not simply look at the packaging, and not listen to the hype of tea sellers.

(1)Tea color observation

Tea color observation (that is, to observe tea with eyes) refers to seeing the color of tea cake or dried tea. The appearance of good ripe *Pu'er* tea is brown and red, and the tea is tender and tight. *Pu'er* tea can be divided into loose tea and compressed tea from the appearance. *Pu'er* loose tea is classified according to its tenderness, with the tenderness from Grade 10 to Grade 1. The higher the Grade is, the more tender the tea is.

First-grade ripe *Pu'er* loose tea **Fifth-grade ripe *Pu'er* loose tea** **Ninth-grade ripe *Pu'er* loose tea**

Yunnan *Pu'er* tea is mainly composed of tea leaves of different grades. The appearance of *Pu'er* tea can be measured by four points. (1) Observe the tenderness and freshness of tea leaves. The more tea sprouts are used, the better *Pu'er* tea is. (2) Observe the texture of *Pu'er* tea. The tighter the texture is, the better *Pu'er* tea is. (3) Observe the velvety of tea. (4) Observe the purity of *Pu'er* tea. With loose tea as raw material, *Pu'er* compressed tea has various kinds of form by autoclaving and numerous colors and varieties. According to the different shapes, there are round cake-shaped Qizi cake tea, brick-shaped *Pu'er* brick tea, bowl-shaped *Pu'er* Tuocha and so on. All kinds of tea are as large as several kilograms, as small as a few grams, and there are hundreds of colors and varieties. In addition to the same endoplasmic characteristics with *Pu'er* loose tea, the main requirements for the shape of *Pu'er* compressed tea are as follows: uniform shape, neat edges, complete edges and corners, clear mold, same thickness and moderate tightness. The color of ripe *Pu'er* tea usually appears to be brown-red or dark brown, while that of raw tea is dark green, brown green, yellow brown, brown and so on, and gradually from green to yellow to brown with the extension of storage time.

The color of ripe *Pu'er* tea liquor

The color of raw *Pu'er* tea liquor

the guests.

(10) Serving the elders and ladies first.

2. Key Elements of *Pu'er* Tea Tasting

The longer it is stored, the more fragrant *Pu'er* tea becomes. *Pu'er* tea is known as "live antique." Mr. Yu Qiuyu, the famous Chinese scholar, described *Pu'er* tea in his book *Pu'er Tea Tasting*, "A group of dark and thick leaves are compressed into a pie-shaped tea. When brewed, I am induldged in sipping with smell of aroma." The Chinese saying also goes, "The aroma is thousands of miles away; the taste is in a cup" is the true portrayal of *Pu'er* tea. This is the charm of *Pu'er* tea made of tea leaves.

In the past hundred years, *Pu'er* tea has been favored by consumers because of its excellent quality. At the same time, the unique flavor of *Pu'er* tea is also related to its natural aging process. After the aging of *Pu'er* tea, it is fractured into tea balls of different sizes and shapes, and then dried naturally in a dry place. According to the transport requirements, the package is put into the basket and transported to other places. Historically, Yunnan is located in the frontier of the motherland. Tea-producing areas are located in the frontier of Yunnan Province, with high mountains and turbulent waters. In ancient times, transportation of tea was extremely difficult. The export of tea relied on the horse and cattle. There were long delays on the mountain roads. Some of the horse teams could only walk twice a year, while the cattle teams could only walk once a year. Tea fluctuated on horseback and cattle back for a long time. The sun, wind, and rain made the contents of tea slowly fermented. As a result, the unique lustre of *Pu'er* tea is brighter. *Pu'er* tea is more neutral, healthy and suitable for the intestines and stomach of Hong Kong people. Most people are fond of drinking. Mr. You Yude, chairman of the Hong Kong-Kowloon Tea Chamber of Commerce, summarized the reasons why Hong Kong people like to drink *Pu'er* tea as "five points" (ten words): strong flavor, brewing resistance, mild temperature, health benefits, and cheapness.

To identify *Pu'er* tea, we must first know its origin. The raw material of *Pu'er* tea is Yunnan large-leaved sun-dried green tea which meets the environmental conditions of *Pu'er* tea producing area, especially the tea produced in famous mountains and introduced in the first section of this part. With good raw materials and good processing, such a combination will produce good *Pu'er* tea. This section elaborates the four key elements of *Pu'er* tea tasting, which are color, aroma, taste, and infusion.

2.1 Color

The first step to appreciate *Pu'er* tea is to observe the color, that is, to see the tea color. Different kinds of tea are presented in different forms, and the direct impression of the external form on the tea is whether the shape is correct, bud and leaf tenderness, tea body color, pressing tightness and whether the tea body falls off, etc. Different tea has a differen color, texture, uniformity, tightness and light conditiont. The second step is to look at the color of the liquor, that is, after tea brewing, according to the brightness and turbidity of the color of the tea liquor, we can also distinguish the quality of the tea. The third step is to look at the infused

Table 4-2 Tea-Water Ratio and Brewing Frequency

Type of *Pu'er* Tea	Tea Ware	Capacity of Tea Ware	Tea Quantity Used	Infusion Frequency
Tender-leaf compressed *Pu'er* tea	Lidded-bowls	150mL	6-7 grams	12-13 times
Strong-leaf compressed *Pu'er* tea	Lidded-bowls or boccaro teaware	150mL	7-8 grams	14-15 times
Strong and old-leaf compressed *Pu'er* tea	Boccaro teaware	150mL	7-8 grams	14-15 times
Loose *Pu'er* tea	Lidded-bowls	150mL	6-7 grams	10-12 times
5-10 years stored *Pu'er* tea	Boccaro teaware	150mL	8-10 grams	15-20 times
Over 10 years stored *Pu'er* tea	Boccaro teaware	150mL	Over 10 grams	Over 20 times

Table 4-2 shows only the recommended tea-water ratio and brewing frequency, which can be slightly increased or decreased according to the number of people, the capacity of teaware and the characteristics of tea.

1.6 Tea Serving

When serving tea, the following etiquette should be followed:

(1) Concentration on brewing and tasting tea.

(2) Good preparation for environment, table, teaware and other devices used for tea-making.

(3) Polite behavior in the whole process of tea brewing and tasting.

(4) Carefully boiling water to avoid water splashing and scalding.

(5) Avoiding tea needles and other pin-shaped devices facing guests.

(6) Re-adding water to the tea kettle in counterclockwise direction for welcoming guests.

(7) Drying the teapot whith a tea towel before serving tea for each cup.

Tea Serving

(8) No touching tea or the edge of cup by hands.

(9) Filling 70% of the tea cup with tea liquor. A full cup of tea implies impoliteness to

Because of the different quanlity grade, year and shape of *Pu'er* tea, the requirements of water temperature during brewing are also different. If the water temperature is not properly controlled, the taste will also be greatly affected. For example, when brewing time-honored tea with lower water temperature, the substances in the tea can not be fully leached, and the aroma and the saturation of the tea liquor can not be fully displayed; if the tea with higher material level is brewed with water of too high temperature, it will not only affect the freshness of the tea, resulting in excessive bitterness and astringency, but also produce the taste of the tea with dull ripeness. Therefore, before brewing, we should fully grasp the characteristics of tea products and regulate the temperature of brewing tea. Generally speaking, when brewing, the water temperature of compressed *Pu'er* tea is higher than that of loose *Pu'er* tea. The water temperature of *Pu'er* tea with lower material grade is higher than that with higher material grade. The water temperature of *Pu'er* tea with longer years is higher than that of shorter years. The water temperature of *Pu'er* ripe tea is slightly higher than that of sun-dried green *Pu'er* tea.

(4) Water Adding

When brewing tea, the technique of water adding will directly affect the quality of tea liquor. In water adding, the first requirement is that the water flow should be stable, neither too slow nor too fast. Secondly, the water flow should not be directly poured into the tea body, but slowly added along the pot side. *Pu'er* tea, because of its special shape, has the steps of "awakening" the nature of tea. When awakening tea, the water flow can be stable and slightly faster, so that the tea pieces can be slightly tumbled, fully involved in the water and finally the nature of tea awakened. After brewed for 3~4 cups, the tea leaves have fully relaxed, and the water adding should be gentle, so that the tea substances can be leached naturally. If the water flow is too fast, the color of the liquor will become muddy. It also affects the taste of tea liquor. When tea leaves are infused to a weak taste, high temperature and rapid water adding can be used to increase the temperature of the tea dregs and accelerate the leaching of the contents.

It should also be noted that when brewing loose or broken tea, as the tea leaf contents will be leached out faster, whether it is to wake up the tea or formally brewed, it is necessary to maintain a gentle water flow, to avoid letting the tea leaves in the pot produce a large gyration. If the water flow is too fast, it will make the tea liquor become cloudy, and also make the tea liquor flavor difficult to control.

1.5 Tea-water ratio and infusions

Tea-water ratio refers to the proportion of water used for making tea. According to the situation and number of people, the tea water ratio varies slightly with the different choice of equipment specifications and brewing tea. Table 4-2 shows the differences between tea-water ratio and infusion according to different characteristics of *Pu'er* tea.

springs in the mountains should be the primary choice, water in rivers should be the second, and water in wells is the last. Less good water can't bring the clean aroma and pure mellow out of tea leaves. Water suitable for tea has to be clear, live, sweet, light, and cool, which are the five criteria for choosing tea water. Clearness means water is pure and transparent. Liveness means water is flowing without bacteria. Lightness requires fewer minerals exist in water. Sweetness and coolness denote water taste ssweet, smooth and cool.

The relationship between water and the acidity and alkalinity of tea liquor

The liquor color is sensitive to the pH value. When the pH is less than 5, it has little influence on the liquor color. If the pH is more than 5, the total color will deepen accordingly. When the pH reaches 7, theaflavins tend to be oxidized automatically and darkened, which results in the loss of the freshness of the liquor taste. Therefore, the pH value of tea-making water should not exceed 7. Neutral water or weak acid water should be used, otherwise the quality of tea liquor will be reduced.

The relationship between tea liquor and soft water and hard water

Tea-making water is usually divided into soft water and hard water. Soft water contains less than 8 mg of Ca and Mg ions in 1 liter of water, and hard water contains more than 8 mg of Ca and Mg ions. Making tea with hard water will affect the taste of tea liquor. So soft water should be used to make tea.

(2) Water Boiling

Pu'er tea needs boiling water to brew in order to make the tea fully stretch, the aroma and taste get the best presentation. Although there is cold infusion method, that is, tea is infused in cold water for more than ten hours to produce flavor, this method is relatively rare, and it is not suitable for compressed *Pu'er* tea. We should learn the method of boiling water before brewing *Pu'er* tea.

The emphasis on boiling water was recorded as early as ancient times. In Tang Dynasty, Tea sage Lu Yu has elaborated the importance of boiling water. Lu Yu said that the first sound of boiling water indicates the best boiling water for brewing tea, and do not use the boiling water for brewing tea when the third sound of boiling is heard. Modern scientific study has shown that mineral ions in water keep changing during boiling. Ca and Mg ions in water will precipitate during boiling process. If boiling time is too short, Ca and Mg ions have not yet precipitated completely, which will affect the taste of tea liquor. Long-boiling "old water", containing trace nitrate at high temperature will be reduced to nitrite. Such water is not conducive to tea making because it is harmful to human health.

(3) Water Temperature

The water temperature of tea brewing is an important factor affecting the extraction of water-soluble inclusions and volatilization of tea aroma. According to the characteristics of different tea, we should master the different temperature of tea making. The water temperature of *Pu'er* tea brewing is generally between 90℃ and 100℃. The temperature below the boiling point referred to here is reached after the water has been boiled and cooled to the desired temperature, rather than using the unboiled water directly for brewing.

Table 4-1   Comparison of Tea Tasting Experience Using Different Kinds of Teawares

Teaware (according to material)	Thermal conductivity	Adsorbability (water absorption, Flavor absorption)	Aroma of tea liquor	Taste of tea liquor	Ideal choice for *Pu'er* tea	Not ideal choice for *Pu'er* tea
Boccaro teaware	Low	best	better	best	Ripe *Pu'er* tea ideal for *Pu'er* tea stored for more than 5 years.	Not ideal for delicate, tender and newly-made *Pu'er* tea.
Porcelain teaware	Medium	better	best	better	Ideal for most *Pu'er* tea.	Not ideal for *Pu'er* tea stored for over 5 years.
Glass teaware	High	good	good	good	Ideal for loose or tender *Pu'er* tea.	Not ideal for most *Pu'er* tea.

1.3 Method of loosening *Pu'er* tea

(1) Methods of loosening the compressed *Pu'er* tea:

Place the compressed *Pu'er* tea in the tea loosening plate.

Open the wrapping paper and put the concave side on the top.

Put left hand on the cake-shaped *Pu'er* tea covering with part of the wrapping paper. Hold the tea needle using right hand.

Insert the needle forward from the concave center of the tea cake and loosen compressed tea into pieces.

Loosening *Pu'er* tea

(2) Dos and dont's

In the whole process of prying tea, be careful of using the tea needle, and do not hurt oneself. Also, do not touch tea leaves directly with hands. What's more, when *Pu'er* tea is compressed, the cake-shaped tea is ususally made by three-layer structure using different tea leaves. To have a thorough tasting of the cake-shaped *Pu'er* tea, one should get tea leaves from the three layers. One also needs to pay attention to the size of the loosened tea piece.

1.4 Water for tea making

(1) Water Choosing

The Chinese saying goes "good water brews good tea". The water quality is extremely important to the quality of tea. Tea Sage Lu Yu pointed out in *The Classic of Tea* that the

(2) Choice of teawares

There are various teawares made of different materials, such as pottery, porcelain and glass. Using different material-made teawares to brew tea results in distinct flavor, aroma and taste of tea. The following part elaborates the features of different teawares.

A. Pottery teaware

The boccaro teaware is the most famous one among pottery teaware, which is made of pottery clay with high iron content. Is is known as as the Best Clay out of Clays. As it should be fired at the temperature between 1,100–1,200℃, the finished product is fine and dense in quality with excellent sand texture. It is believed that the boccaro teaware is the good choice for brewing *Pu'er* tea, as it is proved that the tea leaves in the teapot will not go sour over night even in the hot summer.

There are four features when using the boccaro teawares:

Density: It's ideal for holding the aroma of tea.

Breathability: It's good for keeping tea leaves fresh.

Absorbability: It can absorb the tea liquor and hold the flavor of tea.

Cold and hot resistance: Heat transmission is slow, and can be heated on the fire.

B. Porcelain teaware

Porcelain teaware is a symbol of Chinese civilization. According to the color of the glaze, the porcelain teaware can be classified into four categories: celadon teaware, white-glazed teaware, black-glazed teaware, and multicolor-glazed teaware. Porcelain teaware has good aborbability and medium thermal conductivity. When making tea with porcelain teaware, the aroma of tea is rich and lasting, and the tea liquor is transparent for appreciating.

C. Glass teaware

In ancient China, glass was called "glaze". Although the glaze making technology in China started earlier, it was not until the Tang Dynasty that glazed-teaware was made in China with the cultural exchanges between China and foreign countries. Glass teaware has good thermal conductivity and transparent texture. It is widely used in making delicate green tea. People can not only appreciate the structure of tea in water, but also appreciate the tea liquor. Because *Pu'er* tea is a kind of compacted tea, it requires higher water temperature when brewing. It is not recommended to use glass pot for brewing. But you can choose glass filter and glass tea cup, because glass material does not have water absorption, which will not affect the taste of tea. The use of glass tea cup enables the tea drinker to appreciate the color of tea liquor, and drink tea tenderly and warmly.

(3) Comparison of Tea tasting experience using different teawares

Water heating devices: Iron pots, copper pots, pottery pots with induction cooker etc.

Water heating devices

Tea needle: A tool for loosening the compressed tea.

Tea needle

Tea loosening plate: A plate for holding the compressed tea when loosening tea.

Tea loosening plate

Heating base: The water kettle is placed on it.

Heating base

Tea basin: Used for washing teacups or holding litters.

Tea basin

Filter: A tool for filtering tea dregs .

Filter

Tea cup: A small cup for tea tasting .

Saucer: The instrument for carrying tea cup .

Tea cup and saucer

Tea towel: Used to dry and clean the leftovers of pot, cup, tea table, etc.

Tea towel

Tea ceremony supplementary tools ("Six Treasures of Tea Ceremony" or "Six Gentlemen of Tea Caremong"): tea scoop, tea needle, tea strainer, tea clip and tea caddy (used to store the other five treasures/gentlemen.).

Tea ceremony supplementary tools

Tea lotus (Chahe): Put dry tea for tea appreciation.

Tea lotus (Chahe)

(2) Spiritual environment

The humanistic environment refers to the cultural and social atmosphere influenced by tea-server, tea-drinker and tea art. Firstly, both tea servers and tea guests should dress properly and neatly. Secondly, people involved in the tea-making and tea-tasting should behave elegantly and politely. For example, tea-servers need to have good posture and gesture. Thirdly, people should talk nicely with acceptable pitch, tone and volumn.

1.2 *Pu'er* tea brewing utensils

(1) Brewing utensils

Teapot: The main instrument used to brew tea. It is mainly made of clay, porcelain, glass or metal.

Teapot

Lidded-bowl: Also known as "Sancai Cup" for tea boiling. The implication for "Sancai Cup" is that lid is used to refer to the sky, bowl for people, and saucer for earth. The combination of lid, bowl and saucer connotes the harmony among sky, people and earth.

Lidded-bowls

Justice mug (Gongdao mug): A tea sharing utensil.

Justice mug

Section Three *Pu'er* Tea Brewing and Tasting

1. *Pu'er* Tea Brewing Skills

Tea brewing is the process of leaching the substances of tea with boiling water. Since ancient times, China has paid great attention to tea brewing techniques and accumulated rich experience. The process of making tea requires the coordination of tea, tea sets, water use, environment, brewing skills and so on, so as to highlight the nature of tea.

There are six types of tea in China, and each type of tea has its own features from raw material selection to production technology. To highlight the best flavor of each type of tea, the brewing methods used are also different. Due to its different production technology, *Pu'er* tea can be further divided into raw tea and ripe tea. According to its storage shape, *Pu'er* tea is classified into compressed tea and loose tea. The compressed tea has various shapes such as cake-shaped, brick-shaped, bowl-shaped, and cylinder-shaped. The quality of *Pu'er* tea is determined by its tea leaf selection from different tea gardens and its storage time. All these features make *Pu'er* tea brewing techniques rich and colorful. This part illustrates the brewing techniques of *Pu'er* tea from the aspects of environment, tea utensils, tea-loosening, and water.

1.1 The environment for *Pu'er* tea brewing and tasting

(1) Physical environment

Tea-making and tea-tasting is a process and medium of communication, relaxation, enjoyment and reflection. There are strict requirement for the physical environments of tea-tasting. If people drink tea with too much noise in untidy places, it will have a great impact on the mood of tea-tasting. It will not only make them unable to appreciate the beauty of tea, but also spoil tea. Therefore, the physical setting of tea-tasting is the prerequisite for tea-making and tea-tasting. The tea-tasting environment does not require luxury and grandeur. It requires neatness, quietness and elegance. Under such circumstances, we can calm down, taste the charm of tea and appreciate the essence of tea ceremony.

Xiaohua shows that the aroma of *Pu'er* tea is in doldrums at low temperature, but could be in glamorous lift at opposite temperature. When temperature goes out of the range, it will accelerate the oxidation of tea, stultify the effective substances, and harry the quality of *Pu'er* tea.

2. Humidity

The relative humidity of the storage room should be kept at about 65%, and the increase of humidity can promote the propagation of microorganisms, but when the humidity exceeds 70%, the air humidity will absorb a large number of aroma released from tea, accelerating the release of *Pu'er* tea aroma. However, after more than 80%, the mold of tea grows rapidly, which is easy to cause deterioration and ripening of *Pu'er* tea, resulting in spicy aroma and taste, and turbidity of liquor color.

3. Requirements for Other Storage Conditions

Impurities are easily interlocked with tea, so the storage of *Pu'er* tea is required to be free of peculiar smell. Generally, a special storage room will be efficaciously serving the purpose. Otherwise, light can decompose the effective components of tea, ultraviolet rays will exert adverse effect on the enzyme activities and photochemical reaction will be thus above the surface. Therefore, *Pu'er* tea should be kept away from light, also the appropriate ventilation of storage environment contributes effectively to affecting aging process. Specially speaking, the occasional ventilation, first of all, can blow away the stale and miscellaneous flavors in the tea. Secondly, the environment where ventilatory efficiency is guaranteed can expand contact between tea and oxygen, which is conducive to the proliferation of microorganisms in tea, thus accelerating the changing process of *Pu'er* tea.

In short, in the process of post fermentation or aging of *Pu'er* tea, humidity really matters. Any failure of temperature or humidity control could lead mildew of tea down the drain. Over the course of natural aging, due to the influence of both regional and climatic conditions, in which some auxiliary methods should be employed to make sure the temperature and humidity are better regulated and adjusted. For example, dehumidifiers can be adopted to reduce the humidity in summer in the south while heating and humidification can be of great use to create a suitable storage environment in winter in the north to improve the quality of aging process.

Moldy *Pu'er* tea

Section Two Storage of *Pu'er* Tea

As a typical post-fermentation tea, *Pu'er* tea has the characteristic of "aging becomes more fragrant". *Pu'er* tea, especially raw *Pu'er* tea, is known as a living organism, and its sensory quality and unique flavor can only be formed after a long post-fermentation process.

Aged raw *Pu'er* tea

From the perspective of sensory quality, tagged along with the extension of storage time, the color of raw *Pu'er* tea can drain from light dark green to light brown. On top of that, the bitterness slides in graduality, with throughout sweetness and the liquor color converted simultaneously from shallow brown to amber-straight deep brown. In this duration, the color of cooked tea parches from reddish brown to maroon, the aroma resuscitates to mediating pure, and the taste accentuates to mellow and sweet.

From the point of view of the chemical composition and function of *Pu'er* tea, with the extension of storage time, chances are that the contains of tea polyphenols, catechins, free amino acids, thearubin, theaflavin and soluble sugar retention in both raw green tea and ripe tea would decrease significantly, aligned with increased flavonoids, and tea browns could be ballooned in large quantities. Moreover, the antioxidant activities and NO_2 scavenging approaches of *Pu'er* tea are in high stake of rousing conspicuous and regular enhancement, whereas such mutations are holding strong bearing on the inhibition of α- amylase.

There are two principal factors spiking in the formation of the time-honored *Pu'er* tea: Initially, the selection of raw materials. Only the tea in possess of airtight materials and stable quality can fumble its access to the essence of originality; The second is the decision of storage environment. Due to the influence of external factors such as moisture, oxygen, temperature, light, and microorganisms, the contents will be taking risks of oxidization and polymerization, which will besmirch the color, taste, aroma and other aspects of *Pu'er* tea.

The storage conditions that affect the quality of *Pu'er* tea include the following aspects.

1. Temperature

The temperature with regard to preserving *Pu'er* tea should be maintained at the spectrum between 20℃ and 30℃ (the best temperature is 25℃ ± 3℃) throughout the year since the aging process of *Pu'er* tea is a gradual enzymatic reaction. Below 20℃, on the one hand, the activities of intrinsic enzymes of *Pu'er* tea will be sagging. On the other hand, when the temperature exceeds 50℃, the enzyme protein will denaturate and the enzymatic reaction will be basically near the nadir, which will hurt the changing process of *Pu'er* tea. At the same time, the aroma of time-honored *Pu'er* tea will also face the jitters of apropos temperature. A research conducted by Bao

of cultivation and planting, and historical records show that in 1716, the Deputy Commander in Chief of Kaifa Prefecture paid tribute to Kangxi with "40 cake-shaped *Pu'er* tea and eight baskets of Daughter's tea..." It is the earliest record of tribute tea in Yunnan in the historical materials that have been verified. The tea germplasm resources are abundant, and here is one of the regions with the richest distribution of tea tree species in the world. The local tea water extract is rich in substance and high amino acid. Tea here has the characteristics of "strong aroma, fresh sweetness, and refreshing taste".

According to the survey of ancient tea germplasm resources in Wenshan Prefecture from 2013 to 2017, there were 8 counties (cities) with rich and diverse distribution of tea germplasm resources. The tea named by the preliminary investigation include Guangnan tea, Guangxi tea, Dachang tea, Wuzhu tea, Houzhou tea, Guang e Houzhou tea, MaGuan tea, Zhouye tea, Tufang tea, tea, *Pu'er* tea and Baimao tea, a total of 9 species and 3 variations, accounting for about 35% of the world's tea species. And there are still unidentified species and genera of resources, making it the region with the highest distribution of tea plants in the world. Xichou County is the origin of the Houzhou tea model specimens, and the wild tea species named after the county are "Maguan Tea" and "Guangnan Tea". According to incomplete statistics, the wild tea germplasm resources in the Prefecture are totally 24 tea germplasm resource groups, with a total distribution area about 12,000 hectares, 460,000 plants. According to incomplete statistics, the Prefecture holds cultivated ancient tea tree (garden) area of more than 3,460 hectares, whith the local native Baimao tea as the main species, and a small number of *Pu'er* tea species; cultivated tea tree germplasm resources are mainly distributed in Malipo, Xichou, Maguan, Guangnan, the total distribution area of four counties about 2,500 hectares, and the number of plants is about 570,000. The most representative ancient tea mountains are Malipo ancient tea mountain, Xichou Pingzhai ancient tea mountain, Maguan Gulinjing ancient tea mountain, Wenshan Laojunshan ancient tea mountain, Guangnan Jiulongshan ancient tea mountain, and so on.

width are 17.5 centimeters × 6.6 centimeters. Main features: Leaves are oval, smooth, thick, dark green. Buds are sparse in fuzz and purple in color, with an average flower diameter of 4.8 centimeters. Ovary is hairy, and stigma is 5-lobed.

Brief introduction to tea mountains:

(1) Ancient tea tree group of cultivated type in Huangjiazhai, Mangshui Village, Mangshui Town. At an altitude of 1,840 meters, the distribution area of ancient tea trees is 100 mu, of which more than 400 are relatively concentrated, and the tree age is over 500 years.

(2) Wild ancient tea trees in Baojiawazi, Chashan River, Yanjiang village, Mangshui Town. At 2,348 meters above sea level, there are wild tea groups like "Hongku tea" and "Baohong tea".

(3) Wild ancient tree groups in Yangquan slope, Yanjiang Village, Mangshui Town. At an altitude of 2,340 meters, the ancient tea trees are distributed intensively. There are 20 tea trees with base diameter of more than 40 centimeters and plant spacing of about 3 meters on the ridge of more than 100 meters, which belong to Dali tea.

(4) Wild ancient tea groups in Banana Forest, Lianxi Village, Wenquan Township. The distribution area of ancient tea trees here is large. There are more than 1,000 large tea plants with a diameter of more than 60 centimeters at the base. Among them, the largest tea plant is 15 meters in height, 2.85 meters in stem circumference at the base and 6 meters×6 meters in width. It belongs to Dali tea.

(5) Ancient tea trees in Shifo Mountain, Xinhua Village, Pastoral Town. At 2,140 meters above sea level, the ancient tea tree group has a large distribution area. There are 5 large ancient tea trees, among which the largest one is commonly known as "Willow Leaf Green locally". Its base stem circumference is 3.03 meters, tree height is 14.8 meters, and tree width is 6 meters × 8.4 meters. In July 1997, Professor Liu Qinjin of Southwest Agricultural University went to the field to verify that it was a cultivated ancient tea tree of Dali tea subfamily, and the age of the tree was more than 1,000 years.

(6) Cultivated ancient tea tree in Poshitou Group, Lianxi Village, Wenquan Township. The largest one is 5.8 meters tall, 5.1 meters × 5.4 meters in width, 2.6 meters in diameter at the base with 4 branches. It is the small arbor that spreads out, with green leaves and metersany fuzz. It belongs to *Pu'er* tea, known locally as Yuantou tea, and is one of the witnesses of Yunnan as the origin of tea plants. According to the evaluation of national tea germplasm resources, the tea plant has high content of tea polyphenols and catechins, good quality and high aroma.

5. *Pu'er* Tea Zone in Wenshan District

Wenshan Tea Area is located in Wenshan Prefecture in southeast Yunnan, where the Tropic of Cancer runs through the whole territory. The terrain is high in the northwest and low in the southeast, with the lowest elevation of 106 meters and the highest elevation of 2,991 meters. Most areas are between 1,000 and 1,800 meters above sea level. It is the first front facing the ocean's warm and humid air flowing north to the Yunnan-Guizhou Plateau. The climate is warm and humid, which is very suitable for tea growth. Tea has a long history

4. *Pu'er* Tea Zone in Baoshan District

4.1 Location, environment and climate

Baoshan Tea Zone is located in the west of Yunnan Province, with the terrain high in the north and low in the south. The Lancang River passes through its east. Among the four main tea-producing zones in Yunnan, Baoshan Tea Zone has the highest latitude, the highest average altitude, the lowest temperature and the least rainfall. Baoshan City, Changning, Tengchong, Longling and Shidian all have large-scale tea production. Baoshan has superior natural conditions suitable for the growth of tea plants and has abundant resources of tea varieties. It is an important producing zone of Yunnan "Dianhong" and *Pu'er* tea. From 1986 to 1987, Changning, Tengchong and Longling counties were listed as the first batch of tea base counties for excellence and national export of black tea. There are 100,000 mu of clonal tea base and 150,000 mu of pollution-free tea production base in the city.

4.2 Germplasm resources and distribution of tea mountains

Changning County is located in the middle and lower reaches of the Lancang River, with a relatively high altitude, forming a stereoscopic weather of low heat, warm, cool and cold. The forest coverage reaches 46.7%. There are abundant tea varieties in Changning County. The wild ancient-tree tea planted in the tea garden mainly includes Qing tea and Baohong tea. Found in 1981 in the primary forest of Aganliangzi in Jieshuilu, Qing tea has a height of 700-800 centimeters, a width of 300 centimeters × 500 centimeters, a diameter of 8-10 centimeters, and a leaf area of 71 square centimeters. The leaves are elliptical, green in color, thin and soft in texture, slightly raised in surface, and the buds are wrapped with multiple scales. Baohong Tea belongs to Dali Tea. In the primitive forest of Shizitang Liangzi, which connects Datianba Township with Mangshui Township, the wild ancient-tree tea resources are distributed. Among them, the largest one is 1,000 centimeters tall, 600 centimeters × 700 centimeters in width and 15 centimeters in diameter. The cultivated ancient-tree tea planted in the tea garden is mainly Dali tea, rattan tea and Changning tea.

Among the ancient-tree tea resources in Longling County, there is a tree with a height of 18.2 meters, a width of 5.8 meters, and a trunk diameter of 123 centimeters. It is an arbor with an upright posture. Its leaf length and width are 13.3 centimeters × 6.6 centimeters. The leaves are oval, thick and glossy, dark green, with few hairs on the bud leaves. The flower is large, with an average diameter of 5.8 centimeters, 11 petals, hairy ovary, and 5-lobed stigma.

Among the ancient-tree tea resources in Tengchong County, one tree is 7.7 meters high, 2.5 meters wide, and 29.3 meters in trunk diameter. The tree is arbor with upright posture. Its leaf length and

level, with a relative height difference of 2,979 meters.

Yongde Daxue Mountain, Linxiang Daxue Mountain and Shuangjiang Daxue Mountain constitute the main peaks of the mountain range. Lancang River and Nujiang River are the two major water systems in Lincang, including Luozha River, Xiaoheijiang River, Nanting River, Nanbang River and other rivers. Lincang is under the control of a low-latitude subtropical mountain monsoon with little difference in temperature between the four seasons, in which the dry and wet seasons are distinct. It is extremely favorable to plant growth and tea tree reproduction. Linxiang District, Fengqing County and other 6 counties (districts), 77 townships. 947 administrative villages (including communities) are under the jurisdiction of Lincang City. The total area is 24,500 square kilometers.

3.2 Germplasm resources and distribution of tea mountains

Among the four main tea-producing zones in Yunnan Province, Lincang City is the largest, with a picking area of nearly 900,000 mu, accounting for nearly one third of the Province's total area, ranking first in tea production in Yunnan Province. Lincang has a long history of planting and producing tea. 23 ethnic people are living in Lincang, and they all have the tradition of planting and producing tea. Lincang has 7 counties and 1 district, all of which have large areas of wild tea communities and cultivated ancient tea gardens. Lincang is rich in tea resources. Through investigation, there are 4 tea lines and 8 varieties in the city.

(1) Wild tea: The most representative ones are in Mengku Town, Shuangjiang County; in Big Black Mountain, Heling Village, Nuoliang Town, Cangyuan County; and in Dalang Dam, Manghong Town, Gengma County.

(2) Cultivated ancient tea gardens: There are 23,160 mu of cultivated ancient tea gardens with more than 100 years of age in Lincang City, including 10,600 mu in Fengqing, 5,460 mu in Yunxian County, 5,000 mu in Linxiang District, 2,000 mu in Shuangjiang County, 100 mu in Cangyuan County, Gengma, Yongde and Zhenkang, etc.

The distribution of tea zone in Lincang Distirct is as follows:

(1) Linxiang District: Bangdong, Nahan, Xigui.

(2) Shuangjiang: Pojiao, Bangma, Xiaohusai, Bawai, Nuowu, Nanpo, Bingdao, Laozhai, Dijie, Baka, Bangbing, Banggai, Dadusai, Bangmu, Heigong.

(3) Zhenkang/Yongde: Yongde Daxue Mountain, Mingfeng Mountain, Mangbo, Mangfei, Ma'anshan, Yanzitou, Hanjiazhai, Mengbao, Meiziqing.

(4) Geng Ma/Cangyuan: Hunan, Papo.

(5) Yun County: Baiying Mountain, Hetaoling and Dazhai.

(6) Fengqing: Xiangzhuqing, Pinghe, Chahe, Yongxin.

① Ning'er: Kunlu Mountain, Xinzhai and Ban Mountain.

② Lancang: Jingmai, Mangjing, Bangwei and Pasai.

③ Jiangcheng: Tianfang.

④ Jinggu: Yangta, Kuzhu, Wending Mountain, Huangcaoba, Longtang, Tuanjie.

⑤ Mojiang: Midi, Jingxing.

⑥ Zhenyuan: Qianjiazhai, Laowu Mountain, Tianba, Ma Deng, Mengda and Zhentai.

⑦ Jingdong: Yubi, Jinding, Laocangfude, Manwan.

With large traditional tea production, *Pu'er* Tea Zone has a high historical position. Wuliang Mountain traverses the whole tea zone. There are many kinds of tea, such as the wild tea from Qianjia Village in Zhenyuan County, the large-medium-small mixed leaves in Kunlu Mountain, the big white tea from Yangta of Jinggu, and the large area of rattan tea in Laowu Mountain. In the later period, some new famous tea varieties are planted in the Simao, Ning'er, Ximeng and other places, such as Yunkang No. 10, Xueya No. 100, Ziya, and Zijuan. Most of the *Pu'er* Maocha produced in the *Pu'er* Tea Zone is fragrant and soft, such as the sweetness of Wuliang Mountain Tea and the unique orchid aroma of Jingmai Tea.

2. 3 Features

Traditional sun-dried raw tea in tea zone is mainly made by mixed picking, no moisture evaporation, steam fixation with high temperature, heavy rolling and sunshine exposure. Therefore, these tea strips are black and tight with strong astringency, and are mostly used for fermentation with pilling. In recent years, as *Pu'er* Mountain Tea is becoming popular, the tea zone has been refined and the production process improved. The tea production process in Simao Tea Zone and Jinggu Tea Zone is close to that in Menghai. Jingmai tea, the most popular one, is highly aromatic and astringent because most trees are mixed in the rainforest. The sun-dried raw tea produced by the techniques is symmetrical with exposed fuzz and brightness. The tea liquor is yellow with low astringency and high aroma, where the tea dregs are yellow and moist.

3. *Pu'er* Tea Zone in Lincang District

3.1 Location, environment and climate

Lincang City is located in the southwest of Yunnan Province, between 23°05′ ~ 25°02′ north latitude and 98°40′ ~ 100°34′ east longitude. Lincang is connected to *Pu'er* City in the east, Baoshan City in the west, Dali Bai Autonomous Prefecture in the north, and Myanmar in the south. Lincang City belongs to the southern extension of Nushan Mountain Range of Hengduan Mountain System, which is a longitudinal valley area in western Yunnan. The terrain is high in the middle and low around, and gradually inclines from northeast to southwest. The highest point in the territory is Yongde Snow Mountain at 3,429 meters above sea level, and the lowest point is Qingshui River in Mengding Town at 450 meters above sea

developed in recent years. The tea is oily and bright, with few yellow pieces and black strips, and the new tea liquor is golden and bright, with heavy bitterness, little astringency, and no obvious grass taste. The tea dregs is usually free of scorched flakes and red stalk, bright and tough. After aging for some time, the aroma is better, and the liquor is thicker with quick and lasting sweetness.

2. *Pu'er* Tea Zone in Pu'er District

2.1 Location, environment and climate

Pu'er City is located in the southwest of Yunnan Province, between 22°02′ ~ 24°50′ north latitude and 99°09′ ~ 102°19′ east longitude, bordering Honghe River and Yuxi on the east, Xishuangbanna on the south, Lincang on the northwest, Dali and Chuxiong on the north. It borders Vietnam and Laos on the southeast and Myanmar to the southwest. The border line is about 486 kilometers long (303 kilometers with Myanmar, 116 kilometers with Laos and 67 kilometers with Vietnam). With a total area of 45,385 square kilometers, *Pu'er* is 208.5 kilometers from north to south, 55 kilometers from east to west, and 299 kilometers from south to north. It is the largest prefecture (city) in Yunnan Province. The municipal authorities are stationed in Simao Town, Simao District, 1,302 meters above sea level, 415 kilometers from the provincial capital Kunming by road, 305 kilometers by air route, and 35 minutes by air. There are Bulang, Yao, Hani, Yi, Dai, Wa and Lahu people, and the ethnic customs of *Pu'er* are very different. Due to the influence of subtropical monsoon climate, most of *Pu'er* is frostless all year round with no severe cold or heat. It enjoys the reputation of "Pearl of the Green Sea" and "Natural Oxygen Bar". The altitude of *Pu'er* City is 317~3,370 meters, of which the urban area is 1,302 meters. The average annual temperature of *Pu'er* City is 15~20.3 ℃, the annual frost-free period is more than 315 days, the annual rainfall is 1,100~2,780 mm, and the negative oxygen ion content is above Grade 7.

2.2 History and distribution

According to the Records of *Pu'er* Prefecture in the Qing Dynasty, tea was planted in *Pu'er* Prefecture as early as 1,700 years ago during the Three Kingdoms Period. The earliest person who recorded the cultivation of *Pu'er* tea in historical documents was Fan Chuo, a Tang official who personally visited Nanzhao Prefecture in Yunnan Province (862 A.D.). In his book *Man Shu*, he said: "Tea comes from the mountains around Yinsheng City. Tea is dried and stored after fixation with sunshine. The ethnic people in Nanzhao Prefecture cook the tea with pepper, ginger, cinnamon and drink it." Yinsheng City is Jingdong County of Simao City today, where Yinsheng Jiedu in the Tang Dynasty governed today, Simao City and Xishuangbanna. Historical records show that as early as 1,100 years ago, tea was abundant in Sipu District. Xie Zhaoqi, a scholar of the Ming Dynasty, mentioned the word "*Pu'er* tea" in his book *Dianlue*. The book said: "What all the people drink is *Pu'er* tea, which is steamed into a dough". After 1879 (A.D.) of the Qing Dynasty, France and Britain successively set up customs in Simao, which increased the export of *Pu'er* tea, and the ancient *Pu'er* Tea Horse Road became prosperous. As a cultural relic, many ancient teas and horse roads, ancient post stations and horseshoe prints on the stone record the history of the tea caravan. The main distribution areas of *Pu'er* tea zone are:

divided into 5 climatic zones:

Northern tropical zone. It is the valley area on both banks of Nanju River, Mengwang River and Lancang River, and is lower than 750 meters above sea level.

Southern subtropical zone, warm in summer and warm in winter. It is on both sides of Nanju River with an altitude of 750~1,000 meters.

Southern subtropical zone, warm in summer and cool in winter. It includes Menghai and other places with an altitude of 1,000~1,200 meters.

Southern subtropical zone, cool in summer and warm in winter. It includes Hejian in Meng'a and other places with an altitude of 1,200~1,500 meters.

Central subtropical zone. It is located in 4 townships including Xiding, and Mengman with an altitude of 1,500~2,000 meters.

There are 25 ethnic groups living in Menghai County, including Dai, Hani, Lahu and Bulang. The main distribution is as follows:

① North: Mengsong of Menghai , Huazhuliangzi, Naka, Baotang.

② West: Nanqiao, Bada, Zhanglang, Mannuo, Hesong.

③ East: Pasha, Nannuo (Duoyi, Banpo, etc.), Manmai, Hekai (Bangpeng, Guangbie, etc.).

④ South: Mengsong of Jinghong , Bulang Mountain (old Banzhang, new Banzhang, etc.).

The main tea species in Menghai tea zone is Menghai big-leaf species, which originated from Nannuoshan Mountain. It has plump buds, thick leaves and strong stalks. In Menghai Tea Zone, traditional home-made tea is usually harvested with standard two leaves. There were few large-scale primary processing institutes in the early days, but have been relatively

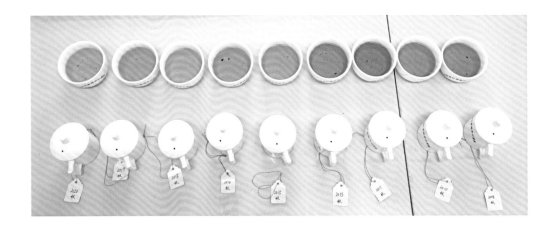

Yiwu, Banna Mengla, Banna Mengpeng and Yiwu Yao Autonomous Regions were established, belonging to Xishuangbanna Autonomous Region (Prefecture). In 1957, They were merged into two county-level banners, Yiwu and Mengla. In 1958, it was merged into Yiwu County and renamed Mengla County in 1959. On March 22, 2002, Mengla Town was set up, with the town government stationed in Manlie Village, and the three villagers' groups in the former Mengla Town were placed under the jurisdiction of Mengban Town. On September 30, 2004, Manla Yi and Yao Township and Mengrun Hani Township were abolished and placed under the jurisdiction of Yiwu Township and Mengpeng Town respectively. The main tea producing areas within Mengla are concentrated in the northern mountains. The six famous ancient tea mountains in history are all in its Yiwu Township and Xiangming Township except You Le. The main tea mountains are distributed as follows:

① Yiwu: Manxiu, Sanqiu Field, Luoshui Cave, Ma Hei, Dingjiazhai, Guafengzhai, Old Street, Wangong.

② Manzhuan: Manzhuang, Manlin and Manqian.

③ Gedeng: Xinsafang, Zhiban, Xinfa.

④ Mangzhi: Jiangxi Bay, Yanglin, Dongjiazhai, Hong tupo.

⑤ Yibang: Mansong, Mangong, Jiabu, Hebian, Malishu.

⑥ Youle: Yanuo, Longpa, Situlaozhai, Mozhuo.

The tea species in Mengla tea area are quite mixed, including wild and transitional ancient trees (variational purple-diameter tea), but mainly large-leaved tea. Compared with Mengla tea, Yiwu tea has larger and longer leaves with less fuzz. There is also a special small-leaved tea called Liutiao tea in Mengla tea area, of which Mansong tea is the most famous one.

(2) Menghai tea zone

Menghai Tea Zone is mainly located in Menghai County. Menghai County can be

1.2 Distribution of Tea Zone

Tea is produced from almost the whole territory of Xishuangbanna. Within Jinghong, there is Youle Ancient Tea Mountain of the six ancient tea mountains (now known as Jinuo Mountain). To the east of Jinghong is Mengla tea zone, which is the source of the famous Ancient Tea-Horse Road in history and the location of the six ancient tea mountains. To the west of Jinghong is the Menghai tea zone, which is the location of the more popular new six tea mountains in recent years (except for Jingmai, which belongs to the Lancang tea zone in *Pu'er*, the other five mountains are in the territory of Menghai). Since the Lancang River passes through Jinghong, people are used to calling the six ancient tea mountains as the tea zone inside the river, and the six new tea mountains as the tea zone outside the river.

(1) Mengla tea zone

"Mengla" is the Dai language, where "Meng" means flat dam or area, "La" means tea, that is, "the place where tea is offered". Legend has it that when Sakyamuni toured here, people offered so much tea that he could not drink it all and poured it into the river. This river is called "Nanla"(that is, "Tea River"), hence named "Mengla". Mengla belonged to Ailao Prefecture of Yizhou County in the Western Han Dynasty, to Jiuliao Prefecture in Yongchang County in the Eastern Han Dynasty, to Pubu Prefecture in the Sui Dynasty, to Yinsheng Jiedu in the Kingdom of Nanzhao of Tang Dynasty, to Jinglong Kingdom in the Song Dynasty. In the seventh year of Chunxi in the Song Dynasty (1180 A.D.), Mengla was governed by the leader of the Dai nationality, Payazhen. In the Yuan Dynasty, it was administrated by the military and civilian administration of Cheli Road, and then by the Cheli Propaganda and Consolation Department in the Ming and Qing dynasties. In the fourth year of Longqing in the Ming Dynasty (1570 A.D.), it was divided into 12 banners, with Mengla and Mengban in Mengla County as one, Mengpeng, Mengrun and Mengman as one, and Zhengdong, Yibang and Yiwu as one. In the seventh year of Emperor Yongzheng's reign in the Qing Dynasty (1729 A.D.), Mengla general manager was set.

In 1913, it belonged to the administrative sub-bureaus of the fifth (Mengla) and sixth (Yiwu) districts of the General Administration of Pusi Border Region. In 1927, the fifth district was changed into Zhenyue County and the sixth district into Xiangming County. In 1929, Xiangming County was incorporated into Zhenyue County, belonging to *Pu'er* Road. On November 6, 1949, the People's Government of Zhenyue County was established under the jurisdiction of Ning'er District. In 1953, Zhenyue County was abolished and Banna

Section One Five Major Zones of *Pu'er* Tea in Yunnan Province

Pu'er tea mountains are widely distributed. From a large regional perspective, Yunnan *Pu'er* tea mountains are mainly distributed in the middle and lower reaches of the Lancang River basin. This book introduces the five *Pu'er* tea zones recognized by the tea industry and tea lovers, and introduces some representative and popular tea mountains.

1. *Pu'er* Tea Zone in Xishuangbanna District

1.1 Location, environment and climate

Xishuangbanna District is located at the southernmost tip of Yunnan, between 21°08′~22°36′ north latitude and 99°56′~101°50′ east longitude. It is connected with Laos and Myanmar, and close to Thailand and Vietnam. It covers nearly 20,000 square kilometers, and the border line is 966 kilometers long. The Mengla Tea zone and the Menghai Tea zone is the typical ones there. Mengla County, where the Mengla Tea zone is situated, is the southernmost part of Yunnan Province, between 21°09′~22°23′ north latitude and 101°05′~101°50′ east longitude. Its location to the south of the Tropic of Cancer. It borders Laos on the east and south, Myanmar to the west, Jinghong to the north-west, and Jiangcheng County in *Pu'er* City on the north, with the national border of 740.8 km (677.8 km between China and Laos and 63 km between China and Myanmar). Mengla County is 868 km away from the provincial capital of Kunming and 172 km from the prefecture capital of Jinghong. In Mengla tea zone, the altitude is 700-1,900 meters, the annual average temperature is 17.2 ℃, and the annual average precipitation is 1,500-1,900 mm. The main ethnic groups include Jinuo, Yi, Hani, and Yao people. The diet is mainly spicy and raw. Architecture is mostly of Dai-style. Rising prices in *Pu'er*h tea have boosted farmers' incomes in recent years, so new-style buildings have been built in virtually every village.

Menghai tea zone belongs to Menghai County. The County is located in the southwestern part of Yunnan Province and the western part of Xishuangbanna Dai Autonomous Prefecture, between 99°56′~100°41′ east longitude and 21°28′~22°28′ north latitude. It is connected to Jinghong City to the east, *Pu'er* City to the northeast, Lancang County to the northwest, and Myanmar on the west and south. The border is 146.6 kilometers long. The longest distance between east and west is 77 km, 115 km between north and south, with a total area of 5,511 square km, of which 93.45% is in mountainous area and 6.55% in the dam area. Menghai Town is 776 kilometers away from Kunming, the provincial capital, and 40 kilometers away from Jinghong, the prefecture capital. Menghai County has a tropical and subtropical southwest monsoon climate, with no severe cold in winter and no extreme heat in summer. The annual temperature difference is small and the daily temperature difference is large. According to the altitude, it can be divided into northern tropical, southern subtropical and mid-subtropical climate zones. The annual average temperature is 18.7℃, the average annual sunshine is 2,088 hours, the average annual rainfall is 1,341 mm, and the frost period is about 32 days. The fog is the characteristic of the Menghai Dam area, with an average annual foggy day of 107.5~160.2 days.

Chapter 4
Drinking and Appreciating *Pu'er* tea

the regulatory effects of the polysaccharides on postprandial blood glucose, fasting blood glucose and antioxidant status of alloxan-induced hyperglycemic mice. The study showed that the content and chemical composition of *Pu'er* tea polysaccharides differed according to the aging time. Five-year-old *Pu'er* tea polysaccharides (PTPS-5) had the highest yield (3.66%), followed by the three-year-old (PTPS-3) (2.24%) and the one-year-old polysaccharides (PTPS-1) had the lowest yield (0.79%). The protein content of the three *Pu'er* tea polysaccharides increased with aging time, and the glucuronic acid contents of PTPS-3 and PTPS-5 were also remarkably higher than that of PTPS-1 ($P<0.05$). GC analysis revealed that although the proportions of their monosaccharide varied, they were all dominated by galactose, arabinose and mannose, along with monosaccharides such as glucose, rhamnose and fucose. According to the molecular weight determination, the aging time could enhance the content of low molecular weight polysaccharides in *Pu'er* tea polysaccharides. *Pu'er* tea polysaccharides boasted strong antioxidant activity and outstanding α-glucosidase inhibition ability, which were closely related to the aging time. Among the four (ABTS radical scavenging capacity, DPPH radical scavenging capacity, FIC ferric ion chelating capacity, FRAP reducing capacity) different antioxidant evaluation systems, PTPS-5 was found to have the strongest ABTS radical scavenging capacity ($IC_{50}=0.49$ mg/mL), DPPH radical scavenging potential ($IC_{50}=1.45$ mg/mL), FRAP reducing ability (FRAP value of 1623.07 at a concentration of 1 mg/mL), and FIC ferrous ion chelating ability ($IC_{50}=0.73$ mg/mL). In addition, PTPS-5 proved to be the strongest a-glucosidase inhibitor ($IC_{50} = 0.063$ mg/mL), being markedly higher than the positive control acarbose ($IC_{50} = 0.18$ mg/mL), whereas PTPS-3 also shared a similar a-glucosidase inhibitory capacity as acarbose (IC50 = 0.19 mg/mL). Both the antioxidant activity under the four antioxidant systems and the strength of inhibition of α-glucosidase were in the following order: PTPS-5 > PTPS-3 > PTPS-1. *Pu'er* tea polysaccharides are positive regulators of the antioxidant status in tetracosan-induced diabetic mice in vivo. The gavage of PTPS-5 at a dose of 40 mg/kg improved the MDA content and SOD activity in the serum and liver tissues to the same level as those of normal mice, even the GSH-Px activity was notably higher than that of normal mice ($P<0.05$), indicating that *Pu'er* tea polysaccharide had a positive effect on the antioxidant status in diabetic mice.

(HDL-C), low-density lipoprotein cholesterol (LDL-C), glutathione peroxidase (GSH-Px), trace malondialdehyde (MDA) and superoxide dismutase (SOD), and to observe the general condition and pathological changes of liver and kidney tissues. In this way, they would find the effects of *Pu'er* tea on the regulation of lipid levels and the protection of vascular endothelial cells in rats with experimental hyperlipidemia. According to the results, medicine, oolong tea and *Pu'er* tea all effectively reduced TCHO, TC, LDL-C and SOD, HDL-C, AST, MDA and GSH-PX in the model rats ($P< 0.05$, $P< 0.01$). Among them, *Pu'er* tea was significantly better than medicine and oolong tea. The study concluded that medicine, oolong tea and *Pu'er* tea could remarkably regulate the body's blood lipid levels, and effectively prevent hyperlipidaemia and oxidation.

Ren et al. gathered the volatile substances of *Pu'er* tea by simultaneous distillation extraction (SDE) to analyse their chemical composition by GC-MS. And the antioxidant activity of the volatile substances at different fermentation stages were evaluated by DPPH and FRAP to determine the relationship between the antioxidant activity and the content of the main components. As a result, the relative content of methoxybenzenes greatly improved during the fermentation. The scavenging capacity of DPPH radicals and the total antioxidant capacity of FRAP of volatile substances showed a positive trend with increasing fermentation level, rising by 100% and 296% respectively after fermentation out of the pile. And the two capacities were positively correlated with the relative content of methoxybenzoids and linalool oxides.

Chen compared the main chemical components of *Pu'er* tea polysaccharides aged for 1, 3 and 5 years and evaluated their in vitro antioxidant properties. He also studied

green tea > black tea > *Pu'er* tea.

SOD and GSH-PX are critical antioxidant enzymes that scavenge free radicals in the body and play a vital role for the balance of oxidation and antioxidation. The findings revealed that both black tea and green tea were effective in increasing SOD activity, with black tea being slightly higher than green tea, while *Pu'er* tea inhibited SOD activity, basically in line with Kuo's report. SOD activity was not detected in the serum of any group in the experiment. It might be that the high-fat ingredients in the feed were more likely to form free radicals such as ROO•, RHOO•, etc. after entering the bloodstream, and a large number of downstream products from the conversion of free radicals inhibited the activity of SOD. All three types of tea had a promoting effect on the activity of GSH-PX, moreover, the effect in liver tissues was stronger in the *Pu'er* tea group than in the other tea groups. MDA is a lipid peroxide formed when oxygen free radicals attack unsaturated fatty acids in biological membranes, which reflects the extent of lipid peroxidation and the damage to cells caused by free radical attacks in the body. It was reported that *Pu'er* tea aqueous extracts had superior ability to prevent lipid peroxidation than green tea and black tea when increased to a certain concentration, which might explain the marked reduction in MDA levels in *Pu'er* tea group compared to the control group, while no significant difference was found for green tea group. Some special health functions of *Pu'er* tea might be closely related to the polyphenols including the oligomers of catechins. We once reported that the ES layer, separated from the ethyl acetate extract layer of *Pu'er* tea, had a stronger ability to remove light free radicals and superoxide anion, in addition, its protective effect on H_2O_2-induced HPF-1 cell damage was more powerful than that of EGCG. This also indicated that some unknown high molecular weight polyphenols, such as catechin derivatives or polymers, deriving from the fermentation of *Pu'er* tea, could have the same or even better antioxidant effects than EGCG.

Pu'er tea boasts a notably complex chemical composition, and it's compounds including polyphenols, flavonoids and polysaccharides are highly antioxidant active. The compounds extracted from ethyl acetate part, in other words, the antioxidant active part, are mainly catechins, flavonoids (kaempferol quercetin and yangonin) and glycosides of flavonoids, all of which show a high number of hydroxyl groups and a strong free radical scavenging capacity. Gallic acid is one of the major antioxidant active ingredients in *Pu'er* tea. Aiming to compare the antioxidant activity of *Pu'er* tea from five Yunnan origins, Jin et al. selected 3-year-old *Pu'er* tea to determine its antioxidant activity and free radical elimination activity by DPPH. The findings suggested that the extracts from the five origins all had certain antioxidant activity, in which the strongest antioxidant capacity was found in the tea produced in Xiaguan, Dali, Yunnan Province, exhibiting an EC_{50} value of 8.88 mg/L, and the weakest one was in Pu'er, Yunnan Province, whose EC_{50} value was 21.81 mg/L. The strength of the antioxidant activity was in the following order: *Pu'er* tea produced in Xiaguan, Dali > *Pu'er* tea produced in Xishuangbanna > *Pu'er* tea produced in Lincang > *Pu'er* tea produced in Honghe > *Pu'er* tea produced in Pu'er. This demonstrated that *Pu'er* tea was an excellent natural antioxidant and free radical eliminator, whose antioxidant activity varied slightly due to different origins.

Jiang et al. established a hyperlipidemic rat model with high-fat diet feeding method. The experimental rats were gavaged with raw *Pu'er* tea, ripe tea, oolong tea and medicine for 35 days to measure the levels of alanine aminotransferase (ALT), aspartate aminotransferase (AST), total cholesterol (TCHO), triglycerides (TG), high-density lipoprotein cholesterol

a small amount of TF3G was detected in green tea and *Pu'er* tea. Theaflavin content in black tea was above that in green tea and *Pu'er* tea. A predominance of EGC and EGCG was found in the catechins of green tea, accounting for about 70% or more of the total. There was a linear relationship ($P<0.01$) between the logarithm of the concentration of each tea powder and the scavenging rate in the DPPH reaction system. The required concentration for 50% DPPH inhibition(IC_{50}) was derived from the linear equation. Results showed that the scavenging effect of each tea powder on DPPH radicals ranked from most effective to least effective in the order of green tea > black tea > *Pu'er* tea. Basically, the body weight of the mice did not change throughout the test period. Although the body weight decreased slightly at the end of the experiment, there was no significant difference ($P>0.05$) from the initial period. Compared with the control group, the MDA content in the serum of mice was significantly reduced in the *Pu'er* tea group ($P<0.05$), while no significant difference ($P>0.05$) was found between both the green tea and black tea groups. MDA levels in mouse liver tissues were reduced in both the black tea group ($P<0.01$) and the *Pu'er* tea group ($P<0.05$), whereas there was no significant difference ($P>0.05$) between the green tea group and the control group. The SOD activity in the liver tissues of mice was improved in the green and black tea groups, reaching a highly significant difference ($P<0.01$) when compared to the control group, while that in the *Pu'er* tea group was inhibited, achieving a highly significant difference ($P<0.01$) in comparison to the control group. In mouse serum, SOD activity was not detected in any of the groups. The GSH-PX activity in mice serum was notably higher in both the green and black tea groups, attaining a highly significant difference ($P<0.01$) compared with the control group, by contrast, there was no significant effect ($P>0.05$) on GSH-PX activity in mouse serum in the *Pu'er* tea group. The results of GSH-PX activity in liver tissues showed that all three tea groups could promote GSH-PX activity in the liver of mice. When compared with the control group, the green tea and black tea groups reached a significant difference ($P<0.05$) and the *Pu'er* tea group attained a highly significant difference ($P<0.01$).

Levels of fermentation affect the composition and content of polyphenols in green tea, black tea and *Pu'er* tea. With the involvement of microorganisms, the changes in the polyphenols of *Pu'er* tea are more complicated and a certain amount of flavonoids are produced under prolonged warm and humid environmental conditions. In this study, black tea (aqueous extracts) contained only about 1% theaflavins, and perhaps most of them were further oxidised and converted to substances such as thearubigins or theabrownins. In *Pu'er* tea, most of the catechins had been oxidised and only a certain amount of GC (about 5.4% of the aqueous extract) were present, with a higher content than in green and black teas. Although the content of polyphenols in *Pu'er* tea was less than that of green tea, the aqueous extracts of *Pu'er* derived by ultrafiltration were analyzed to provide more than 50% (w/w) of high-molecular-weight material (MW>3,000 Daltons), furthermore, the content of gallic acid in *Pu'er* tea was found to be higher than that of green tea.

Studies have shown that the capacity of polyphenolic compounds in tea to clear free radicals had far exceeded that of antioxidants such as vitamin C and E. Available research reports indicated that *Pu'er* tea extracts were superior to green tea, black tea and oolong tea extracts in terms of the ability to scavenge free radicals, inhibit lipopolysaccharide-induced NO production in macrophages and chelate iron ions in the Fenton reaction system. The results revealed that the ability to eliminate DPPH free radicals outside the body is in the order of

to investigate their effect on cholesterol biosynthesis during in vitro tests, as well as their hypolipidemic effect in living animals. It was found that PET could lessen cholesterol biosynthesis in a human hepatocellular carcinomas (HepG2) model system and the inhibition preceded the production of mevalonic acid. The effect of *Pu'er* tea in suppressing cholesterol synthesis was also confirmed in animal tests. It also helped to bring down blood cholesterol, triglyceride and free fatty acid levels and increase the excretion of cholesterol in faeces. Sun et al. suggested that the extracts exhibited extraordinary antioxidant activity, scavenging free radicals and lowering the content of unsaturated fatty acids in LDL to reduce the oxidative sensitivity of LDL.

(3) To chelate metal ions. Duh et al. reported that *Pu'er* tea extracts were effective in chelating metal ions, cleaning up DPPH free radicals and restraining lipopolysaccharide-induced NO production in macrophages. As a strong antioxidant, it was able to scavenge DPPH free radicals and inhibit Cu^{2+}-induced oxidation of low-density lipoprotein (LDL).

2. Antioxidant Efficacies of *Pu'er* Tea

A researcher Dong Fang selected and used *Pu'er* tea powder, as well as green tea powder and black tea powder from Zhejiang Longyou Minghuang Natural Food Development Co., Ltd. to carry out animal experiments. The experiment was set up with control group, green tea group, black tea group and *Pu'er* tea group. The mice were administered by gavage at a dose of 0.9(kg · d) for each tea group while an equivalent amount of control solution (distilled water) was provided by gavage to the control group. The body weight changes were recorded daily for each group. After three-week administration, their blood and liver tissues were taken for the determination of MDA content, SOD activity, GSH-PX activity and in vitro free radicals (DPPH).

To start with, the chemical composition of each tea powder used in this study was identified by the researcher. The polyphenol content of green tea was higher than that of black tea and *Pu'er* tea, while the flavonoid content was lower. Black tea and *Pu'er* tea had equivalent amounts of gallic acid and caffeine, both of which were higher than green tea. Only

Section Three Antioxidant Efficacies of *Pu'er* Tea

1. Antioxidant Mechanism of *Pu'er* Tea

In recent years, the relationship between free radicals and various diseases has been increasingly emphasized, and the development of free radical biomedicine has made the search for efficient and low-toxic free radical scavengers——natural antioxidants——a hot spot for research in biochemistry and medicine. One of the vital contents of modern agriculture in the 21st century is to seek and utilize new bioactive substances in agricultural products. Among them, the research on antioxidant activity is of great importance.

The antioxidant effect is considered to be the most significant mechanism of tea's healthcare and anti-cancer properties. *Pu'er* tea is a special post-fermented tea. In the pile fermentation process of high temperature and high humidity, the polyphenolic components, mainly flavonoids and tea polyphenols, undergo complex chemical reactions including microbial transformation, enzymatic oxidation, non-enzymatic auto-oxidation, as well as degradation and condensation under the action of humidity and heat and microorganisms, resulting in the formation of phenolic components with a more complex chemical structure. For example, the content of active ingredients such as gallic acid and theophylcin in *Pu'er* tea increased remarkably, which greatly enhanced the potential of *Pu'er* tea in antioxidation.

The current research reports indicated that the antioxidant mechanism of *Pu'er* tea is broadly through the following three approaches:

(1) To restrain or directly scavenge the generation of free radicals. Lin et al. revealed that *Pu'er* tea extracts had an excellent ability to scavenge light radicals and curb the generation of nitric oxide radicals. Their researches have shown that *Pu'er* tea extracts could effectively eliminate free radicals in the Fenton reaction system, protecting the DNA superhelical structure and preventing strand breakage. Jie et al. suggested that the ethyl acetate extract layer components and the n-butanol extract layer components of *Pu'er* tea extracts had strong potential to scavenge both DPPH and light free radicals. In addition, studies have reported that *Pu'er* tea extracts had a free radical scavenging effect in Fenton reaction system, and their ability to inhibit NO induced by LPS in macrophages and their iron ions' binding effect were stronger than green tea, black tea and oolong tea extracts. Compared with other teas (green tea, black tea and oolong tea), the inhibitory capacity of 200 μg/mL *Pu'er* tea extracts were not significantly different, but when the concentration increased to 500 μg/mL, their inhibitory capacity was superior to other teas.

(2) To prevent lipid peroxidation. It has been demonstrated by Yang et al. that *Pu'er* tea was effective in lowering serum cholesterol levels but did not change its levels of HDL and triglycerides in animal tests, with a marked increase in the ratio of HDL to total cholesterol and a reduction in the atherosclerotic index. Regarding the increase in liver weight, liver cholesterol levels and triglyceride levels caused by a high cholesterol diet, *Pu'er* tea could only lower liver cholesterol levels and had no apparent effect on liver triglyceride levels. In terms of cholesterol lowering, Xiao et al. took the *Pu'er* tea extracts (PET) as test materials

it. The effects of different doses of ripe *Pu'er* tea on BLA, BUN content and LDH activity in mice were investigated. In comparison to the blank control group, the BLA and BUN contents of mice in all doses showed a tendency to decrease after exercise, while LDH tended to increase. The BLA content of mice in the middle and high-dose groups decreased by 22.1% and 27.8% respectively, compared with the blank control group, showing a highly significant difference ($P<0.01$). The BLA content in the low-dose group was 17.9% lower than that in the blank control group ($P<0.05$). The BUN content in the high-dose group was 17.1% lower than that of the blank control group ($P<0.01$). The BUN content in the medium-dose group was 12.6% lower, which was a significant difference ($P < 0.05$). The LDH activity of mice in the high-dose group increased by 26.3% in contrast with the blank control group ($P<0.01$), and the LDH activity of mice in the middle-dose group increased by 15.2% ($P<0.05$). This suggested that ripe *Pu'er* tea could reduce the levels of BLA and BUN while increase LDH activity after exercise, and the best effect was achieved with high-dose ripe *Pu'er* tea. Studies have shown that the content of MG and LG after exercise was positively related to the anti fatigue effect. The test results revealed that the MG and LG contents of mice in all dose groups of ripe *Pu'er* tea raised after exercise in comparison to the blank control group. The content of MG in the low-dose, medium-dose and high-dose groups was 28.8%, 42.5% and 26.0% higher respectively, displaying a significant difference ($P<0.01$). In terms of LG, its content of mice in the medium and high-dose groups increased by 22.0% and 24.9% respectively, compared with the blank control group, both showing highly significant differences ($P < 0.01$). The content of LG in the low-dose group was 16.7% higher ($P<0.05$). This demonstrated that the content of MG and LG in mice after exercise was remarkably promoted by ripe *Pu'er* tea.

Originating from nature, *Pu'er* tea is of long standing with rich cultural deposits. It plays an important role in people's daily lives for its unique advantages in health care. In comparison with synthetic drugs, it has the advantages of being inexpensive and non-toxic and its daily consumption not only eliminates fatigue but also replenishes other nutrients in the body. Therefore, it is of great significance to further study the anti-fatigue effect of *Pu'er* tea.

It has been shown that BLA, BUN and LDH activity levels were important indicators reflecting the aerobic metabolism ability and fatigue degree of the body. The results of Zhang et al. indicated that the BLA and BUN levels of mice in the low-dose, medium-dose and high-dose groups were significantly lower than those of the blank control group after exercise, while the LDH activity levels of mice in the medium and high-dose groups were greatly higher, suggesting that ripe *Pu'er* tea may enhance the LDH activity and remove excessive lactic acid from muscles, thus reducing lactic acid during exercise. With the decrease of urea nitrogen, the adaptability of the body to the load was improved, leading to a delay in fatigue, especially in the high-dose group.

long time, many scholars have expected to find a remedy that is safe, effective and non-toxic to delay fatigue and accelerate its elimination. And tea has the function of "refreshing brain, relieving fatigue, brightening the eyes, promoting urination, eliminating summer heat and clearing heat", showing a promising prospect. Zhang et al. used a representative sample of ripe *Pu'er* tea to explore its anti-fatigue effect through the animal model of mice. The selected three test ripe *Pu'er* tea samples that were produced by Menghai Yuncha Technology Co., Ltd., Yunnan Longrun Tea Group and Yunnan Simao Tea Tree Breeding Farm were mixed in equal quantities as test materials. Extracted by boiling water, the tea samples were filtered by suction filter and combined with the extracted tea liquor. They will be concentrated under reduced pressure, and then be bottled and sterilized to obtain 1g/mL tea liquor concentrate.

The mice were randomly divided into 4 dose groups according to their body weight: blank control group and ripe *Pu'er* tea groups of low-dose, medium-dose and high-dose. With 20 mice in each group, they are half male and half female. The experiment used Kunming mice at 5 times the recommended human dose as the lowest dose, and 10 and 20 times dose for the middle and high doses respectively, that is, 0.5 g/kg, 1.0 g/kg and 2.0 g/kg of ripe *Pu'er* tea for low-dose, medium-dose and high-dose groups respectively. The blank control group was given normal saline (0.9%). The mice were weighed once a week and given oral gavage in the morning (9:00-11:00am) once a day according to their body weight for 30 days. Researchers performed weight-bearing swimming tests in mice, and measured their physiological and biochemical indicators of blood lactic acid (BLA), blood urea nitrogen (BUN), blood lactate dehydrogenase (LDH), liver glycogen (LG) and muscle glycogen (MG). The results showed that the mean body weight of mice in the low-dose, medium-dose and high-dose groups of ripe *Pu'er* tea increased by 25.22%, 28.03% and 18.17% respectively at the end of the experiment compared to the beginning, while that of the negative control group increased by 31.37%. Visually, the mice in the high-dose group were leaner and longer. This indicated that the three doses of ripe *Pu'er* tea all had significant inhibitory effects on body weight gain in mice, and the best effect was seen at high-dose. The duration of weight-bearing swimming in mice was a direct response to the anti-fatigue effect, which was positively related to the anti fatigue effect. In contrast with the blank control group, the weight-bearing swimming time of mice in the three doses groups of ripe *Pu'er* tea was prolonged, with an increase of 25.74%, 51.27% and 56.00% respectively. Among them, the weight-bearing swimming time of the middle and high-dose groups was significantly different from that of the blank control group ($P<0.01$). It showed that the ripe *Pu'er* tea in the medium-dose and high-dose groups could extremely prolong the weight-bearing swimming time. It has been reported that the levels of BLA and BUN in the body were negatively correlated with the anti-fatigue effect, while LDH activity was positively correlated with

Section Two Anti-Fatigue and Anti-Aging Efficacies of *Pu'er* Tea

1. The Concept of Fatigue and Aging

The concept of health changes as our society makes progress and the medical model alters. Anti-fatigue and anti-aging have become a concern. Clinically, fatigue is a common manifestation of sub-health. However, it is a very ordinary symptom or phenomenon, which can not only exist in healthy people and sub-healthy people, but also in many people with diseases. There are roughly 3 reasons: continuous work that exceeds the body's ability, some negative emotions and diseases. Being a subjective feeling of reduced function of local tissues and organs or general discomfort caused by various reasons, fatigue can be divided into transient fatigue and cumulative fatigue. It can be manifested physically, for example, a sense of decreasing physical strength and weakness; or mentally, in the form of a feeling of aversion to activity (physical or mental). Its behavioural impact can be seen in the decline of work efficiency.

Senility, known as aging, refers to the degenerative changes that occur in various tissues and organs of the organism as we grow older. It is the comprehensive manifestation of many physiological and pathological processes and biochemical reactions of the organism, as well as the joint effect of internal and external factors (including heredity, nutrition, mental factors, emotional changes, environmental pollution, etc.). Aging is an inevitable trend in the development of human life. It is an objective law that is not subject to human will. No one can prevent the process, but it can be slowed down by scientific methods.

The external characteristics of human aging mainly include: (1) The skin becomes saggy and wrinkled, especially on the forehead and around the eyes. (2) The hair gradually turns white and thinner. (3) Senile plaques appear. (4) Dental bones begin to atrophy and fall out. (5) Bones becomes loose and brittle. (6) Gonad and muscle become atrophied, showing various symptoms of "menopause", such as menstrual disorders and weight gain in women, as well as depression, hyperactivity and insomnia in men. (7) There is also vascular sclerosis, in particular of the cardiovascular and cerebral vessels, and the degeneration of the elastic tissue in the lungs and bronchial tubes.

The major functional characteristics include: (1) People's eyesight and hearing are diminishing. (2) Their memory and thinking skills gradually decrease. (3) They will have slower reaction and movement as well as lower adaptability. (4) They will also encounter reduced cardiopulmonary function and metabolic dysfunction. (5) Decreased immunity will make them vulnerable to germs, which may even lead to autoimmune diseases. (6) There will be age-related diseases such as hypertension, cardiovascular disease, emphysema, bronchitis, diabetes, cancer, prostate enlargement and geriatric psychosis.

2. Anti-Fatigue and Anti-Aging Efficacies of *Pu'er* Tea

Chronic fatigue has become a disease that plagues people's normal work and life. For a

($P<0.01$). Before the treatment of *Pu'er* tea, the TC levels of rats in the obesity model group and experimental treatment groups were consistent. After treatment, the levels of TC in the three dose groups of *Pu'er* tea declined visibly compared with the obese model group, reaching a significant level ($P<0.05$). The decrease in TC levels of rats in the diet control group showed a highly significant difference ($P<0.01$). At the same time, relative to the obesity model group, the serum TG level of the rats in the middle and high dose treatment groups of *Pu'er* tea reduced by 20.10% and 25.62% respectively. The serum TG levels of rats in the *Pu'er* tea high-dose treatment group had approached the TG levels of the diet control group (27.21%). These experimental results show that both *Pu'er* tea and diet control can effectively lower the serum TC and TG levels in nutritionally obese rats, and can improve the quality of serum indexes in rats, which have potential value for the prevention or treatment of obesity in rats. Moreover, the effect of high-dose treatment is close to that of diet control, which will provide a way for those who purposely control their diet for weight loss to use *Pu'er* tea as an alternative.

Being a major component of total serum cholesterol, high-density lipoprotein cholesterol (HDL-C) is considered as the "good cholesterol" in animals. Xiong Changyun's experiments demonstrated that both *Pu'er* tea and dietary control treatment were effective in increasing the serum HDL-C level in nutritionally obese rats compared with the obesity model group. After treated with *Pu'er* tea for 6 weeks, the content of serum HDL-C in obese rats in low, medium and high-dose treatment groups increased by 27.59% ($P<0.05$), 43.68% ($P<0.01$) and 62.07% ($P<0.01$) respectively, while that in diet control group only increased by 21.84% ($P<0.05$). It indicated that the effect of *Pu'er* tea on the improvement of HDL-C in obese rats was better than that of dietary control alone, showing a dose effect, that is, the highest dose *Pu'er* tea treatment group was found to exhibit the best effect. And what was more encouraging was that the level of HDL-C in the highest-dose group far exceeded that of the blank control group, reaching a highly significant difference level ($P<0.01$).

Developed by the international medical community, atherosclerosis index (AI) is used to measure the degree of atherosclerosis. The AI values of obese rats that treated with different doses of *Pu'er tea* and diet control went down evidently. Compared with the obese model group, the decrease was 57.38%, 69.20% and 79.75% respectively in the three dose treatment groups, while diet control group was 62.87%, which reflected a highly significant difference ($P<0.01$). By comparison, the AI values of the diet control group were equal to those of the blank control group, indicating that the diet control could restore the degree of atherosclerosis to the previous level. While the AI values of the medium and high-dose treatment groups were lower than those of the blank control one, and the high-dose *Pu'er* tea treatment group reached a highly significant difference in particular. Such results show that *Pu'er* tea has a prominent effect in anti-atherosclerosis. It can not only inhibit the increase of AI value in obese rats caused by excessive intake of high-fat diet, but also improve the serum indicators of normal rats, reducing AI value and lowering the risk of atherosclerosis, which is beyond the reach of simple diet control.

activity of antioxidant enzymes SOD was higher than that of normal control group. Domestic researchers have also demonstrated that sun-dried raw tea or *Pu'er* tea fed to mice on a high-fat diet were both effective in inhibiting the elevation of blood lipids, and could reduce the levels of serum TG, TC, LDL-C to the normal range in an all-round way, while significantly increasing the level of high-density HDL-C. And the effect of *Pu'er* tea was better than that of sun-dried raw tea.

By using basic animal feed (M02-F) and high-fat feed (M04-F), Xiong Changyun mixed M04-F with 2.5%, 5%, and 7.5% of ripe *Pu'er* tea powder to form three doses of tea-containing high-fat feed at low, medium, and high dosages, and fed them to the test rats after mixing in a blender. Through the comparative experimental study among blank control group, obesity model control group, diet control group, *Pu'er* tea low-dose group, *Pu'er* tea medium dose group and *Pu'er* tea high-dose group, it was found that the body weight of the rats in the diet control group tended to decrease significantly in relation to the obesity model group, reaching a significant difference ($P<0.05$) at the end of the experiments. Among the three dose treatment groups of *Pu'er* tea, the weight of rats in the middle and high-dose groups also decreased obviously, and their effects were close to those of the diet control group, while the low-dose *Pu'er* tea group did not show any inhibitory effect. On the other hand, interestingly, the food intake of rats in each treatment group did not show great difference, indicating that the weight gain of obese rats was not reduced by food intake after treatment with different doses of *Pu'er* tea.

Serum total cholesterol (TC) and triglyceride (TG) levels are important indicators for the evaluation of obesity in rats. At the end of the experiment, the serum TC and TG levels of rats in the obesity model group were remarkably higher than those in the blank control group

of cerebral embolism and heart failure in obese people is twice than that of normal weight people, and the incidence of coronary heart disease, hypertension, diabetes and cholelithiasis is 3 ~ 5 times higher. People's life expectancy will be significantly shortened due to the attack of these diseases. Tending to be afraid of heat, obese people are sweaty and their skin folds are susceptible to dermatitis and abrasions, which are prone to purulent or fungal infections. Meanwhile, the increase in body weight leads to a heavier burden on various organs, rendering them vulnerable to various kinds of traumas, fractures and sprains, and so on. In addition, it is also directly related to sleep apnea syndrome and the development of malignant tumours.

2. The Efficacies of *Pu'er* Tea on Obesity

As for the weight loss effect of *Pu'er* tea, the earliest researcher was Japanese scholar Mitsuaki Sano, who proved in 1985 that feeding *Pu'er* tea to high-fat rats could reduce the content of cholesterol and triglycerides in the blood vessels of high-fat rats, which significantly reduced their abdominal adipose tissue weight. Subsequently, Yang also reported in 1997 that after feeding *Pu'er* tea, the rats dealt with high cholesterol molding had reduced food and water consumption, lost weight, decreased cholesterol and triglyceride content in blood and liver, while high-density lipoprotein cholesterol content was increased. In 2005, Kuo et al. showed that after 30 weeks of feeding *Pu'er* tea, the weight, cholesterol and triglyceride contents of normal rats were significantly reduced, and the reduction was greater than that of other teas, such as green tea, oolong tea and black tea. Meanwhile, low-density lipoprotein cholesterol was reduced, while high-density lipoprotein was significantly increased, and the

Section One Lipid-Lowering and Weight-Losing Efficacies of *Pu'er* tea

It has been confirmed by scientific research that *Pu'er* tea is effective in reducing fat and losing weight, providing cosmetic and life-prolonging benefits. The growing number of drinkers has brought it into the limelight as the "slimming tea" and "longevity tea". Nowadays, the consumption of *Pu'er* tea has changed its traditional role of quenching thirst. According to its characteristics, the tea is considered to be a medicine that is associated with health, longevity, fat reduction and weight loss, which adds new connotation to it.

1. The Overview of Obesity

1.1 The Concept of Obesity

Adiposis, known as obesity disease, is a nutritional and metabolic disorder caused by genetic factors and environmental effects. The most obvious feature is that the body takes in more energy than it consumes, resulting in excessive accumulation of fat in the body. At present, it has become a global public health concern. According to the International Obesity Task Force, it will become the biggest killer threatening human health.

1.2 The Characteristics of Obesity

Individual obesity mainly shows increases in the number and the volume of fat cells. Generally speaking, it is called obesity when a person's weight exceeds more than 20% of the standard weight, or the body mass index (body mass index, BMI) is greater than 24 kilograms/square meter.

1.3 The Causes of Obesity

Although there are many reasons for obesity, the most basic one is the imbalance of energy metabolism in the body. Many factors can lead to energy metabolism disorders (imbalance), such as excessive nutrition, reduced physical activity, endocrine metabolism disorders, hypothalamus damage, genetic factors or emotional disorders.

1.4 The Hazards of Obesity

Obesity not only affects physical appearance, but is also harmful to health. It will bring about many chronic diseases, for instance, multiple metabolic abnormalities related to cardiovascular diseases, and will increase the morbidity and mortality of Type II diabetes, coronary heart disease, hypertension, stroke, congestive heart failure, dyslipidemia, sleep apnea syndrome, and some cancers (such as ovarian cancer, thymus cancer, and colon cancer). It is the enemy of people's health and longevity. Scientists have found that the incidence

Chapter 3
Health Benefits of *Pu'er* Tea

contents. There is a great difference in the taste of sun-dried green tea produced in different tea mountains and regions. This difference not only reflects the different content of aromatic substances in tea leaves, but also reflects the subtle difference in the contents. How to judge such differences, and how to effectively and reasonably "reorganize" and "blend" such differences to create a better quality *Pu'er* tea product is the dream that *Pu'er* tea people have always pursued from ancient times to the present. The blending of *Pu'er* tea can form a graded transformation of the subsequent fermentation of *Pu'er* tea. Taking the post-fermentation of cake tea as an example, it requires that the compressed cake tea should be moderately tight, that is, not too dense (over-compressed) and not too loose (too large a gap). The best way to solve this problem is to mix different grades of raw materials reasonably, with seven-grade tea as the "skeleton" and three-grade or five-grade tea as the "supplement". The construction of this reticular skeleton can make *Pu'er* tea appear hierarchical, and make the subsequent fermentation appear graded transformation. Many people have a misunderstanding about the grade of *Pu'er* tea, believing that the higher the grade of the selected raw materials, the better, and some people even pursue cake tea made of pure buds (special grade), and mistakenly believe that the higher the grade of raw materials for *Pu'er* tea, the higher the nutritional value, and vice versa, the lower the grade, the fewer nutrients. Taking the content of reduced total sugar in *Pu'er* tea raw materials as an example, the result of testing by authoritative departments is that the seventh-grade tea has the highest content. This is the main reason why seventh-grade tea is used in large quantities for cake tea. This kind of coarse and old tea is not only the main force to form a "reticular skeleton" of cake tea, but also the "backbone force" for the subsequent fermentation and transformation of *Pu'er* tea because of the characteristics of the contained substances.

The blending of *Pu'er* tea is a highly personalized art. All classic *Pu'er* tea products, whether they are old teas that have been passed down for decades or new products in recent years, have their own unique "tea nature". As long as we deeply understand them, we will find their differences, even if they are subtle. This kind of differentiation has a strong personality color, which will make our taste have a deep memory and will not be forgotten for a long time. This kind of feeling, or quality, cannot be endowed by simple "pure material" and "one mouthful material". It is more of a crystallization of the *Pu'er* tea maker's wisdom from long years of experience and enlightenment, and is the "fruit" of superior blending skills.

Yue Cheng Nadameng tea garden in Menghai County

The fourth kind is pure material of the same tea garden, different batches, the tree tea in the same tea garden, divided into different ages and batches. This classification is much more rigorous than the previous, such pure material must be picked at about the same time and the same age of the trees, so such pure material is relatively small.

The fifth kind of pure material, "single plant" is also called "one tree". It means "single plant tea", that is, tea on a single ancient tree. Generally, they choose the oldest tea tree or several tea trees in the ancient tea garden, and only pick its first spring tea, so the quantity is very small. Therefore, it is a private collection, and little is sold in the market. This kind of tea is the favorite of tea enthusiasts.

4.2 Blended *Pu'er* Tea

The "blending" of *Pu'er* tea includes 6 aspects: blending of grades, different tea mountains, different varieties, different seasons, different years, and different degrees of fermentation. Grade blending of *Pu'er* tea is the most common method in blending *Pu'er* tea. Whether it is new tea or old tea passed down with a long year, it will be found that there are differences in the tea grade at the bottom, surface and inside. Even if the difference is small, there are unique blending techniques in it. The "sense of hierarchy" of a cake of *Pu'er* tea is inseparable from the skills of hierarchical blending. Therefore, the blending of *Pu'er* tea has been widespread since ancient times.

The blending of *Pu'er* tea focuses on the "complementary advantages" of the tea's

is the basic rule for traditional *Pu'er* tea production; another view is that the blending of *Pu'er* tea is a technological means to optimize and improve the quality. The classic *Pu'er* tea products must have its own unique "secret formula", and the core content of this formula is "blending". Here are some aspects to clarify how to correctly understand the pure ingredients and blending of *Pu'er* tea.

4.1 Pure *Pu'er* tea

At present, there are five main forms of identification and classification of pure tea in the market:

The first kind is pure material of old trees, Ancient Tree Tea from different mountains. The feature of this "pure material" tea is that it is pure tea as long as it is Ancient Tree Tea, no matter which mountain it is from, or whether it is spring tea or autumn tea.

The second kind is seasonal pure material, the tea of the same season. This pure material is ont divided by the age of the tea tree, nor by batches. It is only divided by seasons, that is, it is divided into spring tea, summer tea and autumn tea.

The third kind is pure material of the same tea garden, the same batch, the same old tree tea in the same tea garden, regardless of batches and ages of the tea trees. The characteristic of this kind of pure tea is being cultivated in the same piece of tea garden, not the same mountain or stockade, but it is not picked in batches.

(1) Small-Tree Tea: Small tree tea refers to the tea tree with a planting life of less than 30 years.

(2) Medium-Tree Tea: Medium tree tea refers to the tea tree with high yield and dense planting with a planting age of 30 to 60 years.

Medium tree tea

(3) Old-Tree Tea: Old tree tea refers to the tea tree with high yield and dense planting with a planting age of 60 to 100 years.

(4) Ancient-Tree Tea: Ancient tree tea refers to ancient trees with more than a hundred years. Ancient tree tea acquires the deep mineral content of the soil and can reflect the uniqueness of each hilltop in the best state with rich endoplasm. *Pu'er* tea made of such raw materials is sought after by "tea enthusiasts". The raw material of ancient tea is limited, and the price on the market is higher, but its unique endoplasm can better reflect the "tea culture" of *Pu'er* tea. In addition, tea trees are classified into cultivated tea trees and wild tea trees depending on whether they are cultivated artificially or not. Currently, tea trees are mainly cultivated tea trees on the market.

Old tree tea

3.5 Classification By Storage Method

(1) Dry Storage *Pu'er* Tea: Dry storage *Pu'er* tea is stored in a clean warehouse environment which is ventilated, drying and in low air humidity (usually referring to air humidity less than 70%), so that the tea leaves can naturally ferment.

Ancient tree tea

(2) Wet-Storage *Pu'er* Tea: Usually placed in high temperature and humidity, the fermentation rate can be sped up. For "wet storage" tea, we should first eliminate the misunderstanding of "wet-storage tea is old tea." Although wet storage tea is aging faster, it will inevitably be moldy under high temperature and high humidity conditions. No matter from the perspective of health or health care, the moldy tea visible to the naked eye, the pungent tea with a sense of smell, and the locking-throat tea with stinging, stabbing and hanging feeling in the throat, tongue and mouth when tasting tea run counter to the requirements of green, pollution-free, ecological and organic food advocated in the new century, so we do not advocate the sale and drinking of wet-stored *Pu'er* tea.

4. The Differences between Pure *Pu'er* Tea and Blended *Pu'er* Tea

Pu'er tea has always been controversial about "pure material" and "blending". One point of view is that the selection of raw materials for "pure material" or "a mouthful of materials"

compression is loose tea. According to its quality characteristics, *Pu'er* loose tea can be divided into 11 grades: super grade, first grade to tenth grade.

3.2 *Pu'er* Compressed Tea

(1) Cake tea: flat-shaped discs, of which each seven-cake tea has a net weight of 357 grams, each 7 for a tube, each tube weights 2,500 grams, hence the name of seven cakes.

(2) Bowl tea: the shape is generally the size of the bowl. There are 100 grams, 250 grams, etc., as well as mini bowl tea, and each has a net weight of 2 to 5 grams.

(3) Brick tea: rectangular or square, mostly 250 to 1,000 grams, this shape is mainly for convenient transportation.

(4) Golden gourd tribute tea: also known as Tuan tea, Rentou tribute tea, is a unique and special compressed *Pu'er* tea. It is named as golden gourd because its shape is like a pumpkin and tea buds are golden after aging. In the early years, Golden gourd tea was specially made for the imperial court, hence the name "Gourd Tribute Tea", ranging from 100 grams to several hundred catties. *Pu'er* tea can also be customized into various shapes according to personal preferences in the market.

3.3 Classification by Processing Method & Techiques

The general production process of *Pu'er* tea: The fresh leaves of Yunnan big-leaf species have undergone the process of fixation, rolling, sunnning, drying and become sunned green tea. With the different processing methods for green tea, *Pu'er* tea has been divided into two series of raw *Pu'er* tea and ripe *Pu'er* tea.

(1) Raw *Pu'er* tea: Raw *Pu'er* tea is a compressed tea formed from the fresh leaves of the Yunnan big-leaf tea tree. Growing within the protection area of geographical indications, it takes sunned green tea as the raw material processed by fixation, rolling, drying and autoclave molding. It has the guality characteristics of dark green color, pure and long-lasting aroma, strong flavor and sweet taste, green and yellow liquor, thick and yellow-green leaves.

(2) Ripe *Pu'er* tea: Ripe *Pu'er* tea is a loose tea and a compressed tea formed from fresh leaves of Yunnan big-leaf species growing within the protection area of geographical indication, and is processed by special technology (piling, etc.) through fermentation (Mircrobial solid-state fermentation). It has the quality characteristics of red and brown appearance, rich red and bright endoplasmic liquor, unique aroma, strong flavor and sweet taste, red and brown leaves.

3.4 Classification By Tea Tree Age

The tea tree species of *Pu'er* tea in Yunnan are mainly arbor tea. According to the different ages of tea trees, *Pu'er* tea can be divided into small-tree tea, medium tree tea, old tree tea and Ancient Tree Tea.

Small tree tea

(2) Unique Tea Tree Varieties: The raw material of tea is the fresh leaves of Yunnan big-leaf tea tree. The big-leaf tea is different from medium-leaf and small-leaf tea. It is a tree in appearance, with larger leaves than medium and small leaves, and the tea buds are thick and tender, rich in contents, high in active ingredients and strong in health care.

(C) Unique Production Process: *Pu'er* tea is made of leaves of Yunnan big-leaf species and undergoes post-fermentation. The production process is: fresh leaves of YCamellia sinensis var.assamica go through the process of fixation → rolling → sunning → sunned green fuzz tea (autoclaved into raw *Pu'er* tea) → piling → turning pile → drying → sifting → picking → mixing (loose ripe *Pu'er* tea) → autoclaved into ripe tea → drying → *Pu'er* ripe tea (compressed tea).

(4) Unique Shape: In addition to loose tea, compressed *Pu'er* tea has various shapes including small shapes such as 3 grams of mini bowl-shaped tea, 100 grams of bowl-shaped tea, 250 grams of bowl-shaped tea, brick tea, cake tea; large shapes such as golden melon, gourd, screen, and large plaque.

(5) Unique Quality: *Pu'er* tea belongs to post-fermented tea and has the characteristic of getting more and more fragrant. The longer it is stored, the better the quality and the more expensive it is, the more it is loved and sought after by old tea people. If properly stored, it can be kept for several decades. Therefore, it is known as an "antique tea" that can be drunk. Nothing but *Pu'er* tea is regarded as an antique.

(6) Unique Drinking & Brewing Methods: *Pu'er* tea is the tea that pays most attention to brewing techniques and drinking art. In Yunnan, the brewing of *Pu'er* tea is very particular. There are researches on water, utensils, etc which will be introduced in detail in the following section on tasting. Except for clear drinking, people like to combine *Pu'er* tea with their own national traditions and customs to form a variety of drinking methods and make various kinds of blended tea.

It is precisely because of the above characteristics that *Pu'er* tea has become an "antique" that can be collected and appreciated. In addition to the red wine from Bordeaux in France, *Pu'er* tea from Yunnan in China is also an "antique" that can be drunk, which is a feature that no tea can have.

3. The Classification of *Pu'er* Tea

Pu'er tea is like a book that can never be overturned. Every time a page is flipped, people always have a confused "noun". The reason is that *Pu'er* tea market has too many words about *Pu'er* tea. "Arbor tea," "big tree tea," "Ancient Tree Tea," "wild storage tea," "dry storage tea," "bowl-shaped tea," and so on, why? Carefully sorting and combing, understanding from the perspective of classification, and grasping its rules, we can not only clarify the clues, but also benefit from seeing through falsehood and return to the essence.

3.1 Classification by Appearance

Pu'er Loose Tea: In the process of tea making, *Pu'er* tea with loose-leaf shape without

Section Two Basic Knowledge of *Pu'er* Tea

1. The Concept of "*Pu'er* Tea"

In 2008, the General Administration of Quality Supervision, Inspection and Quarantine of the People's Republic of China and the Standardization Administration of the People's Republic of China jointly issued the National Standards of the People's Republic of China "*Product of Geographical Indication – Pu'er tea*" (GB/T 22111—2008), and officially implemented on December 1st. At present, the definitions of *Pu'er* tea is generally based on GB/T 22111—2008. The Standards has made clearly normative requirements on the range of origin, tea tree varieties, processing technology, quality characteristics and classification of *Pu'er* tea: taking sun-dried green tea with Camellia sinensis *var.*assamica within the protection scope of geographical indications as raw material, and using specific processing technology within the protection scope of geographical indications to make tea with unique quality characteristics. According to its processing technology and quality characteristics, *Pu'er* tea is divided into raw *Pu'er* tea and ripe *Pu'er* tea. According to its appearance, *Pu'er* tea is divided into loose tea (raw *Pu'er* tea) and compressed tea (raw *Pu'er* tea and ripe *Pu'er* tea).

The Standards stipulates that the protection scope of geographical indications of *Pu'er* tea products includes 11 prefetures (cities) as well as 639 townships under 75 counties (cities) in Yunnan Province, including *Pu'er* City, Xishuangbanna Dai Autonomous Prefecture, Lincang City, Kunming City, Dali Bai Autonomous Prefecture, Baoshan City, Dehong Dai and Jingpo Autonomous Prefecture, Chuxiong Yi Autonomous Prefecture, Honghe Hani and Yi Autonomous Prefecture, Yuxi City, and Wenshan Zhuang and Miao Autonomous Prefecture. Only the tea with the specific raw materials, and produced and processed within the protection scope of geographical indications can be called *Pu'er* tea. In addition, the standards also defines the post-fermentation process of *Pu'er* tea for the first time: the procedure of forming the unique quality characteristics of *Pu'er* tea (raw tea) through a series of transformations of its contents under specific environmental conditions by the combined effects of microorganisms, enzymes, heat and humidity and oxidization of the Yunnan sun-dried green tea of big-leaf species or *Pu'er* tea (raw tea).

In this book, *Pu'er* tea is divided into "raw *Pu'er* tea" or "ripe *Pu'er* tea", abbreviated as "raw tea" and "ripe tea".

2. The Characteristics of *Pu'er* Tea

Pu'er tea, as a special tea, has the distinct characteristics from other teas, namely the "six uniquenesses" of *Pu'er* tea:

(1) Unique Origin: Named after the place and developed into a special tea, *Pu'er* tea has obvious geographical characteristics. The origin of *Pu'er* tea is in the major tea mountains along the lower reaches of Lancang River, especially the six major tea mauntaina. Tea beyond this specific area cannot be called *Pu'er* tea, strictly speaking.

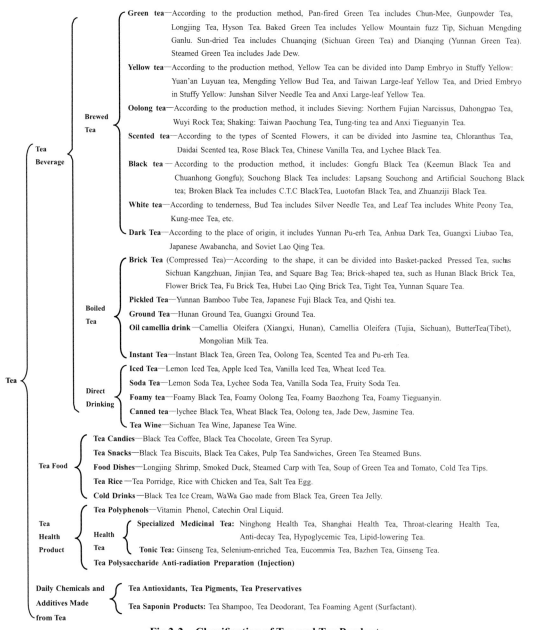

Fig 2-2　Classification of Tea and Tea Products

while brick tea and cake tea belong to compressed tea, and extracted tea is also called instant tea.

The classification method that divides tea into basic tea and reprocessed tea has been adopted in China and the specific classification information is in Fig. 2-1.

Fig 2-1 Comprehensive Classification of Tea

2. The Three-Sphere Integrated Classification

In the 1990s, with the continuous improvement of tea deep processing technology, new tea products emerged one after another. In view of this situation, Professor Liu Qinjin, a tea expert at Southwest University, proposed a Three-Sphere Integrated Classification method for tea, which divides tea into tea beverages, tea food, tea health products, tea daily chemical products and additives. Among them, tea beverages are further divided into three categories: brewing, boiling, and direct drinking.

(2) Black Tea

Black tea belongs to fully-fermented tea. Its production process includes withering, rolling, fermentation and drying. There are three kinds of black tea on the market currently: Souchong black tea, Kungfu black tea and broken black tea.

Black tea

(3) Oolong Tea

Oolong tea is semi-fermented tea. Being the most noted tea in Fujian, Guangdong and Taiwan provinces, it can be classified into Minnan Oolong (such as Tie Guanyin, Huangdan tea, Benshan tea, Meizhan tea, and Maoxie tea), Minbei Oolong (such as Wuyi rock tea, Narcissus tea, Dahongpao tea and Cinnamon Tea), Kwangtung Oolong (such as Fenghuang Dangcong tea, Fenghuang narcissus tea), Taiwan Oolong tea (such as Dongding Oolong Tea, Wenshan Pouchong and Bai Hau Oolong).

Oolong tea

(4) White Tea

White tea is a kind of lightly-fermented tea. Fresh leaves are slightly withered and dried directly in the sun after picked. The main producing areas are in Fujian and Yunnan, such as Pekoe silver needle, white peony, Shou Mei, and Yunnan moonlight white tea.

White tea

(5) Yellow Tea

Yellow tea belongs to the partially-fermented tea. It needs heaping for yellowing to make tea leaves and tea liquor become yellow in the process of production. It can be divided into Huangya tea such as Junshan Yinzhen tea, Mengding Huangya tea and Huang Xiaocha and Huang Dacha.

Yellow tea

(6) Dark Tea

Dark tea is the post-fermented tea that will go through the processes of fixation, rolling, piling, drying, etc. It includes Yunnan *Pu'er* tea, Hunan Anhua black tea, Hubei old green tea, Sichuan Kang Brick tea, Guangxi Liubao tea and etc.

Reprocesset tea mainly includes scented tea, compressed tea and extracted tea. Jasmine tea, chloranthus tea and rose tea belong to flower tea,

Dark tea

Section One Classification of Tea

Tea leaves are made of the buds, leaves and tender stems of tea trees (*Camellia sinensis* (L.) O. Kunts) and are processed by specific techniques without any additives for human drinking or eating. China is the birthplace of tea. It is also the country with the most tea types and the most complete tea production. There are varieties of tea and complex names. There are many classification methods of tea in China, including sales classifiction, production quality or season classification. Professor Chen Chuan, a tea expert in Anhui Agricultural University, proposed that tea could be classified into six types and established the six tea classification systems. In addition, researcher Cheng Qikun from TRI CAAS (Tea Research Institute, Chinese Academy of Agricultural Sciences) proposed a comprehensive tea classification method, which classifies the tea into two basic categories: basic tea (namley the six major types of tea) and reprocessed tea (scented tea, compressed tea, tea beverages, etc.). The method is accepted at home and abroad. In this book, a new method of tea classification is added: Three-Sphere Integrated Classification.

1. Comprehensive Classification of Tea

This method is based on the production process, product characteristics, tea tree varieties, the raw materials of fresh leaves and production areas. Generally, tea is divided into two parts: basic tea and re-processed tea. According to different manufacturing methods and qualities, tea is usually divided into green tea, black tea, Oolong tea, white tea, yellow tea and dark tea.

(1) Green Tea

Green tea belongs to non-fermented tea that is made through fixation, rolling and drying after picked. According to fixation and drying methods, it can be divided into: pan-fried green tea, such as West Lake Longjing tea, and Biluochun tea; steamed green tea, such as Japan's Sencha and Hubei's Gyokuro; baked green tea, such as Huangshan Maofeng tea, Taiping Kowkui tea; sun-dried green tea, such as Yunnan green and Szechwan tea.

Green tea

Chapter 2
Basic Knowledge of *Pu'er* Tea

4. Tea Drinking in Russia

Russians first encountered tea in 1638. Vassili Starkov, a Russian aristocrat, was sent as envoy to the Altyn Khan. Starkov brought some sable fur as a present to one of the Mongolian Khan. In return, he was given four poods (Russian unit, about 64 kg) of tea to Tsar Michael I as a gift. The Tsar began to love the drink after tasting it. Tea came to become a popular beverage for the aristocratic family in Russia.

Since the 1770s, merchants in Moscow have started to import tea from China. During the reign of Emperor Kangxi of the Qing Dynasty, China and Russia signed an agreement on the long-term import of tea from China by Russia. However, the route was a long journey and the transportation was difficult with limited quantity. Therefore, tea became a typical "urban luxury drink" in Russia from the 17th to the 18th centuries. It was not until the end of the eighteenth century that the tea market was expanded from Moscow to a few provinces. By the early 19th century, the trend of tea drinking began to flourish in all aspects of life in Russia.

Russians love drinking black tea with sugar, lemon slices, jam, butter, or milk in order to make it filled with flavor of fruit and sweets, removing the astringency in black tea.

Russian samovars make tea-time distinctive tradition. In Russia, the samovar is seen as symbol of warm home, family ties, and something that is truly Russian. As a saying goes, "It cannot be regarded as tea drinking without samovar." Russian samovars are metal pots used for heating water to make tea. They contain a vertical pipe that heats water and keeps it hot for hours. It is a two-part vessel consisting of a large metal container for water crowned with a teapot with a spigot near its base. In traditional samovars, there was a little stove at the bottom for burning coal to heat up and boil the water. Today, though, samovars rely solely on electric heaters for boiling water. Modern people use a tea pot to brew tea for 3 to 5 minutes by adding water boiled by samovars.

Nowadays, samovars have been mostly replaced with electrical kettles, but they still have a strong presence as a souvenir item that is displayed in a prominent place in a home. There are still those who prefer to use traditionally heated samovars to serve tea for family members and friends, especially during the festivals. Russian samovars remain one of the most well-known symbols of Russia around the world.

and she helped spread tea culture to the upper echelons of society. Gradually, tea has been spread to the public, and become a popular beverage in the United Kingdom.

"When the clock strikes four, everything stops for tea." This British lyric best illustrated the role of tea drinking played in the Britain. According to statistics, the United Kingdom consumes nearly 200,000 tons of tea annually, accounting for 20% of the world's total tea trade. 80% of British people have the habit of drinking tea every day, and each person consume sabout 3 kg of tea every year. British people's tea consumption accounts for almost half of the total tea beverage consumption. This shows that Britain is truly a "kingdom of tea drinking."

The British initially drank green tea and *Wuyi* tea, and later gradually started to drink black tea and developed their own unique tea culture. British people have a lot of tea time every day: sleeping tea, breakfast tea, lunch tea, afternoon tea, and so on. They take time to drink a cup of tea during busy working hours. Tea has become an indispensable and important element of their life.

It is said that Anne Russell pioneered the idea of afternoon tea as a meal and a social event in 1840. Luncheon was a light meal taken around noon and dinner wasn't served until 8 o'clock at night leaving a long gap between the two meals. People would become rather peckish around four o'clock in the afternoon and asked that tea, bread and butter and cake be brought to serve with tea. The afternoon snack gave people a much-needed energy boost and kept them going until dinner. It is said that afternoon tea is one of the important ways of British social activities and is a vital bridge of interpersonal contacts.

even the participants themselves. The final principle, tranquility is the ultimate goal of the tea ceremony. After observing the first three principles, it is hoped that participants will arrive at a state of inner peace and calm. "Harmony, respect, purity, and tranquility" was put forward by Sen No Rikyu (born in 1522), a monumental figure in the history of the Japanese tea ceremony, and has been the code of conduct of Japanese people related to tea for more than 400 years.

2. South Korean Tea Ceremony

South Korea established its tea drinking style from the period of Silla, and it has a history of thousands years. The beginning of tea drinking history and the prevalence of tea culture in Korea have a very close relationship with China.

According to Volume X *The Memorabilia of Silla of Memorabilia* of *The Three Kingdom*s，the envoy who came back from China (the Tang Dynasty) brought tea seeds, and the King ordered him to plant them in the Jirisan mountain. Tea has been planted since the reign of Queen Seondeok, and it became prosperous. In the period of Silla, tea drinking was popular in temples and the royal family, and there was some distinction between tea set and tea ware. Tea drinking and tea rituals were in their peak in Goryeo Dynasty. "Tea house" was set up by the court, and tea was prepared when tea house officials presided over the temple worship, officials welcomed foreign envoys, and king's tour. The Saheonbu set a "tea time", and civil servants would have a clear mind after drinking tea, which ensured efficiency. They advocated that tea drinking is a way that purifies the soul of officials and develops their virtues. Monks used tea in a variety of Budda-worship ceremonies and in their daily spiritual practice life. What's more, tea drinking was also prevalent in public, and there were tea houses and tea shops selling tea.

In the Joseon Dynasty, there was a decline in tea drinking while its culture was undertaking a transition. Scholars in this period of time wrote a number of books pertaining to tea culture. Zen master Venerable Cho-ui was such a saint. He built the "Yichi-an" thatched cottage, around which he planted tea trees, and used the water dropping from of stalactites to brew tea. He wrote the book *Ode to Korean Tea*, which is often compared with *The Classic of Tea* wrote by Lu Yu. Zen master Venerable Cho-ui is considered as the Saint of Tea in Korea by later generations.

The spirit of Korean tea ceremony is moderation (*Chiang Kai-shek* in Korean). In the *Ode to Korean Tea* written by Zen master Venerable Cho-ui, the phrase was mentioned twice. "The water and the spirit of tea may be perfect, it still may beyond the Chiang Kai-shek (moderation), and if Chiang Kai-shek is kept, the tea will be good and full of anima." In 1979, "Chiang Kai-shek" officially became the core idea of Korean tea ceremony culture and a code of conduct for people related to tea at the tea party in South Korean Tea Gathering. "Chiang Kai-shek" contains four meanings: neither being more nor being less, treating all equally, putting others before oneself, seeking truth and returning to nature.

3. Afternoon Tea in Britain

In 1662, Charles II of England married Portuguese Princess Catherine Braganza, and she brought tea with her as part of her dowry. Tea was Catherine's preferred beverage of all time,

Section Four Tea Custom in Various Countries

Tea, coffee and cocoa are the three major beverages. More than 50 countries in the world are growing tea and more than 160 countries have the habit of drinking tea. Tea is not only a way of life for people in the world, but also an important medium for international exchanges. Nowadays, people from different countries meet together to discuss the history and current status of tea in the context of tea and culture exchange. Friendship, understanding and further cooperation have been boosted because of the sharing of tea culture. Tea and tea culture have been undertaking a diverse and dynamic change. Different countries have their own unique way of drinking tea and tea culture. Tea plays an important role in people's life.

1. Japanese Tea Ceremony

The development of Japanese tea history and culture has a very close relationship with China, especially after the Taika Reforms. Japan fully absorbed the cultural system of the Tang Dynasty, and tea and tea-drinking habits were also introduced to Japan.

The current Japanese tea ceremony generally refers to Matcha tea ceremony and Sencha tea ceremony. There are many differences between Matcha tea ceremony and Sencha tea ceremony in terms of tea varieties, teawares, brewing and drinking methods. Ceremonial matcha preparation begins by sifting the matcha powder (powdered green tea) into the bowl. Once the hot water has been added to the powder, we are ready to continue the most well-known step of how to make ceremonial matcha guide, the whisking. We can start by gently stirring the matcha powder into the water, making sure it is all mixed in. With foam on top, the match a paste's ready to drink. Sencha (green leaf tea) is a traditional Japanese tea ceremony focused on loose leaf sencha tea, different from Matcha ceremony which uses powdered green tea. The green tea leaves, which have been dried but not fermented, are brewed in a small pot according to a process designed to bring out their best flavor. The ceremony itself is a beautiful and meditative experience, where every step is carefully performed with grace and precision. From the selection of the tea leaves to the brewing process, each element is executed with meticulous attention to detail.

The Japanese tea ceremony emphasizes harmony, respect, and tranquility, and is performed meticulously for a serene and meditative atmosphere. In other words, "Harmony, respect, purity, and tranquility" are the principles of Japanese tea ceremony. Harmony is the core tenet informing every tea ceremony aspect. In practical terms, harmony is manifested in the balance and unity between the participants, the natural surroundings, and even the utensils used in the ceremony. The principle of respect manifests itself in the conscious actions of the host and guests. Each gesture expresses mutual respect and consideration, from how the host prepares the tea to how the guests receive it. The ritualistic nature of these actions is meant to elevate everyday manners to the level of deep, conscious respect for others and the moment at hand. Purity is reflected in the meticulous cleanliness of the tea room, the utensils, and

the weight of 6.5 *Liang*. There appeared an unfair trading in tea market.

Since the founding of the People's Republic of China, tea production and business has become standardized and more prosperous. In the early 1970, *Qizi* cake-shaped tea was adopted as a brand in order to further promote *Pu'er* tea.

Tea and tea culture are the contribution of Chinese people to the world. Currently, black tea and green tea are widely accepted as a beverage by people all over the world. The scented tea has been spread to many places with the enchanting song *What A Beautiful Jasmine*. *Pu'er tea* in China will be valued by its umami flavor and health benefits.

It was until 1973 that the processing technique for fermented or ripe *Pu'er* tea became successful in Yunnan Province. According to *The Annals of Yunnan Tea Import and Export Company,* in 1973, the company sent delegates to Guangdong to study the processing technique for making ripe *Pu'er* tea. When coming back, the group of people succeeded in producing ripe/fermented *Pu'er* tea after several experiments. It turned out that the processing technique for making ripe *Pu'er* tea by piling was adopted and has been used until today.

4. The Origin of Yunnan *Qizi* Cake-Shaped Tea

Qi, meaning seven, is considered as an auspicious number by Chinese people. *Qizi*, meaning seven sons, is a symbol of many children and blessings in Chinese culture. The beginning and regulation of *Qizi* cake-shaped tea originated in the Qing Dynasty. *The Record of Laws and Systems of the Qing Dynasty* noted that, " In line with the criteria set up during the 13th year of the Yongzheng's reign (1735 A.D.), Yunnan tea should be sold in the quantity of a bundle with seven compressed tea cakes. This seven-compressed tea cakes should weigh 49 *Liang* (about 1.8 kilograms). 32 bundles is equal to 1 *Yin* with the weight of 5 kilograms. Tax differs pertaining to weight." Qing government stipulated that the tea sold to Xizang Province should be in the quantity of a bundle with seven compressed tea cakes. But *Qizi* cake-shaped tea was not mentioned at that time.

At the end of the Qing Dynasty, the shapes of compressed tea were more open and inclusive. For example, Baosen Tea House produced *Pu'er* tea in the quantity of five compressed tea cakes. In order to make a distinction, the seven compressed tea cakes were named as *Qizi Yuan Tea* (Yuan means round in Chinese). But *Qizi Yuan Tea* was not the brand name. During the early period of the Republic of China, there was a chaos regarding the weight of *Pu'er* tea as a commodity. Some tea businessmen tried to reach an agreement on the weight of *Pu'er* tea. Businessmen in Simao County (Today's *Pu'er* City) proposed that each compressed cake should not exceed the weight of 6 *Liang*. While others such as Leiyongfeng Tea Houses produced *Pu'er* tea in the quantity of eight bundled compressed tea cakes, each tea cake with

Record of Yunnan, "All people are drinking Pu tea in the shape of cake made by steaming fresh tea leaves" This was the first written record of Pu Tea. During the Ming Dynasty, numerous caravans shuttled between Yunnan and Xizhang, resulting in the flourish of "tea-and-horse market" in the center of *Pu'er* City. Many cities appeared along the Ancient Tea-Horse Road.

During the middle of the Qing Dynasty, the ancient "Six Great Tea Mountains" were flourishing. People valued *Pu'er* tea, and *Pu'er* tea was used as a tribute to the emperor. In the meantime, *Pu'er* tea was sold to Sichuan, Xizang and Southeast Asia. The route of *Pu'er* tea transaction is knowned as the Ancient Tea-Horse Road.

3. The Historical Development of *Pu'er* tea Processing Techniques

The name "*Pu'er*" stems from its place of origin—the *Pu'er* City of Yunnan Province, China. For thousands of years, *Pu'er* tea has enjoyed a world-renowned reputation for its unique flavor. Beginning in the Eastern Han Dynasty, *Pu'er* tea became popular as a kind of goods and beverage in the Tang and Song dynasties. *Pu'er* tea got its universal name in the Ming dynasty, and was flourishing in the Qing Dynasty. In the seventh year of the Yongzheng's reign (1729) of the Qing Dynasty, the *Pu'er* Prefecture was established and had jurisdiction over the today's *Pu'er* and Xishuangbanna districts. During the Qing Dynasty, *Pu'er* tea was offered to the emperor as a tribute, which greatly fostered the spread of *Pu'er* tea. Historically, due to traffic inconvenience, *Pu'er* tea had to rely on caravans for business through the Ancient Tea Horse Road. For the sake of transportation convenience, tea was pressed into the shape of ball, brick or cake. What's more, In the course of transportation and storage, natural fermentation would happen, forming a unique flavor and health benefits.

During the early 20th century, the production and trading center of *Pu'er* tea was Yiwu area and *Pu'er* area surrounding the ancient Six Great Tea Mountains. Because of the war, tea houses went out of business successively in the 1930s and 1940s, and their production also went into a recession. In May 1938, assigned by the government, Mr. Fan Hejun and a group of technicians established "Fohai Tea Factory of China Tea Company" in Fohai County (currently Menghai County), symbolizing the beginning of tea production by machines. The turmoil also resulted in the prosperous business of tea production in Fohai County.

After the founding of the People's Republic of China, the production of *Pu'er* tea was in the process of recovery and innovation. In 1952, Fohai Tea Factory resumed business, and was renamed Menghai Tea Factory in 1953, which became the leading company of Yunnan *Pu'er* tea production. Before the three major tea factories in Kunming, Menghai and Xiaguang cities began to compress tea leaves. Most of the tea exported to Hong Kong in the early 1950s was loose tea, which has not been fermented. People in Hong Kong couldn't bear the bitter and strong flavor of the raw large tea leaves. There were also some owners of Yunnan tea houses who settled in Hong Kong and brought the traditional method of fermenting *Pu'er* tea. Consequently, the traditional fermented *Pu'er* tea became popular in Hong Kong, and soon was adapted by people in Guangdong Province. Liantonglong, Hengruixiang, Nanji, Shengji, Linji, Baotai, Tong'an Changzhoufuhua were some of the famous tea houses in Hong Kong at that time. It was recorded that in the late 1950s, Yunnan tried to produce fermented *Pu'er* tea by steaming, which turned to be a failure.

Section Three The History of *Pu'er* Tea

Pu'er, also spelled as Pu'erh, is highly prized for its cultural significance, complex flavor, and long history. Let's delve into the historical change and development of *Pu'er* tea.

1. The Origin of the Name of *Pu'er* Tea

In 1729, the Emperor Yongzheng in the Qing Dynasty set up *Pu'er* Prefecture, and its government was located in today's Ning'er County. *Pu'er* Prefecture had jurisdiction over today's *Pu'er* City, Xishuangbanna district and parts of Lincang City. The word "*Pu'er*" comes from Hani language spoken by Hani ethnic people in the area. In Hani language, "Pu" means village, and "Er" means the bay riverside. It turned out that "*Pu'er*" means the village by the riverside, implying home for people. The word "*Pu'er*" was originally used to refer to people living in *Pu'er*, and they were the ancestors of today's Bulang and Wa ethnic people. The study showed that *Pu'er* was first used to refer to people in the area, followed by the place named *Pu'er*. *Pu'er* tea got the name because *Pu'er* tea was planted, produced, and traded by *Pu'er* people living in *Pu'er* area.

2. The Historical Changes of *Pu'er* Tea

In the third year of Xian Tong of Tang Dynasty (862 A.D.), Fan Chuo was assigned to be an official in Yunnan. In the seventh volume of his book *Mangshu* (*The Book of Yunnan*), he wrote, "Tea tree is planted in the mountains of Yingsheng City, and is picked freshly. Mengsheman people drink it by cooking and mixing with beans, ginger and cinnamon." The mountains of Yingsheng City included other tea mountains in Lincang, Dali, Dehong, Honghe, Yuxi, and Baoshan counties. It was credibly recorded that tea in Yingsheng City belongs to Yunnan "large-leaf" species, namely, tea species for making *Pu'er*. Therefore, tea produced in Yingsheng City is supposed to the ancestor of *Pu'er* tea.

Li Shi in the Song Dynasty also recorded in his book Xu Bo Wu Zhi (*The Continuation of Encyclopediia*), "Tea is planted in the mountains of Yingsheng City. The local people drink it by cooking with beans and ginger." Tea drinking and culture became popular in the Tang Dynasty, and was prosperous in the Song Dynasty. The prosperity of Chinese tea is not only a trend followed by many people, but also has promoted the trade and exchange with Zang ethnic people living in Xizang by the Ancient Tea Horse Road.

During the Yuan Dynasty, there was a place called "步日部" (Bu Ri Bu), which was later written as "普耳" (*Pu'er*, there was no "氵" before "耳"). This is how the name "普洱" (*Pu'er*) came into being, and it has been written in the history since then. Tea from Yunnan named as "Pu tea" was transported to Xizang and Xinjiang. "Pu Cha" (Pu tea) gradually became an indispensable commodity in the markets of Xizang and Xinjiang. It was around the end of the Ming Dynasty that "Pu Cha" was renamed as "*Pu'er* Cha" (*Pu'er* tea).

During the Wanli Period (1620) of the Ming Dynasty, Xie Zhaozhe wrote in his *A Brief*

from China.

2.3 Spread Northward to Russia

In 1618, the Chinese envoy brought a few boxes of tea to the Czar in Russia. Due to the long distances, it took about 16 to 18 months for tea leaves to reach Russian consumers. It was until 1903 when the railway running through Siberia was completed that the caravan trade, which lasted more than 200 years, ended. The railway enabled China's tea, silk and porcelain to be shipped directly to Russia within two weeks.

2.4 Spread Southward to India and Sri Lanka

Tea plantation on a large scale was introduced in 1788 in India. In 1824 a tea plant (*Camellia Sinesis*) was brought from China by the British and planted in the Royal Botanical Gardens, Peradeniya, Ceylon.

At present, tea trade is ubiquitous among hundreds of countries and regions on the five continents. More than 50 countries have planted and produced tea, and people in over 160 countries and regions have the habit of drinking tea with a population of over 2 billion. The spread of tea has promoted the development of human civilization and serves as an important bridge and carrier for exchanges between China and other countries. Tea industry and tea culture have also been enhanced globally with the spread of Chinese tea. At the opening ceremony of the 2008 Olympic Games in Beijing, there was a long scroll showing 5,000 years of Chinese civilization with two Chinese characters, namely " 茶 " (Cha, tea) and the other was " 和 " (He, harmony). These two Chinese characters show the essences of Chinese tea culture, which are honor, beauty, harmony and respect (cultivating morality, being honest and money saving, in order to conduct oneself in society harmoniously and honestly, and to respect and love people). It connotes that people all over the world pursue one dream, a dream of peaceful and harmonious world.

brick. During the long journey to Xinjiang and Xizang, the green tea was fully fermented, leading to the change of color from green to dark. Black tea was crafted in the late Ming and early Qing dynasties in the 17th century. The processing techniques used in the creation of black tea is another key determinant of its final taste, scent, and visual appeal. Black tea processing generally encompasses several stages, such as withering, rolling, oxidation, and drying. After withering and rolling, tea leaves will be processed by oxidation (fermentation). During this phase, the enzymes react with atmospheric oxygen to convert polyphenols into new compounds such as catechins and flavonoids. These new compounds bestow black tea with its signature taste. Xiaozhong Black Tea was first crafted by people living in Wuyishan City, Fujian Province. More black tea varieties were added in the coming time, such as Gongfu Black Tea and Hongsui Black Tea. Qing Cha, also known as Oolong tea, has the reputation of green leaves mixed with a reddish hue red, due to the shaking processing skill (tea leaves being oxidized inside bamboo tumblers). Oxidation, besides changing the flavor of the tea, also changes the color of the leaves to a darker hue. Over time, these types of oolong tea have spread to Guangdong and Taiwan. White tea is the most minimally processed of the true teas. The tea production process involves only harvesting, withering, and drying the leaves. Each leaf is picked by hand and then withered for 72 hours either in direct sunlight or in a room with a carefully controlled climate. The leaves are then dried to prevent oxidation from occurring. White tea is rich in amino acids, especially in L-theanine. It is of relatively high demand in international trades.

2. The Spread of Chinese Tea Abroad

Tea planting and tea drinking in the world have direct or indirect relationship with Chinese tea. Tea has been spread domestically from the original place to the reaches of Yangtze River and South China. Moreover, tea has been spread internationally to all over the world, such as Korea, Japan and Russia. In ancient times, Chinese tea was transported by horse and camel through the Ancient Silk Road. It was also transported to other countries by trade routes such as sea.

2.1 Spread Eastward to Japan

As early as the 6th to the 7th century A.D., there were plenty of exchanges of Buddhism between Japan and China. *Saicho*, a Japanese monk, was considered as the person who first took tea seeds into Japan. He returned to Japan after studying in China, and planted tea seeds in the Japanese Hiyoshi Shrine, which then became the most ancient tea garden in Japan. In the early 12th century A.D., when Japanese monk *Eisai* visited China, he not only brought back more tea seeds and Chinese customs of drinking powdered green tea, but also brought back his own understanding of Buddhist teachings. Therefore, tea and Buddhist ideas in Japan were interdependent and eventually developed into a complex and unique ceremony which has been preserved so far, namely, the Japanese tea ceremony.

2.2 Spread Westward to Europe

In 1517, Portuguese sailors brought tea leaves to Portugal from China. In 1560, Portuguese missionaries introduced Chinese tea varieties and tea-drinking methods to Europe. In the early 19th century, afternoon tea was developed in Britain by the Duchess Anna and has prevailed so far. With the increasing consumption of tea, Britain came to import a lot of tea

the Song Dynasty, tea was spread throughout China, and the coverage of the tea production was about the same as that of the modern times. Local officials in Zhejiang Province has the record of providing tea as tribute for emperor. From the early years of the Five-Dynasty Period and the Song Dynasty, the climate of the whole country turned from warm to cold, resulting in the rapid cultivation and production of tea in the South China. Jian An in Fujian Province has become the center for the production of Chinese Tuan Cha (ball-shaped tea) and Bing Cha (cake-shaped tea), leading to the rise of tea production in Southern Fujian Province and the South of the Five Ridges. Before the Ming and the Qing dynasties, the processing technique of tea was basically quite simple, and green tea was the major production. It was during the Ming and the Qing dynasties that tea processing techniques were enriched, resulting in the production of six categories of tea (Green Tea, Yellow Tea, Dark Tea, Black Tea, Oolong Tea, and White Tea). The Ming and the Qing dynasties were the golden ages of tea processing. Currently, China boasts four major tea production areas, which are the north of the Yangtze River, the south of the Yangtze River, South China and Southwest China covering 20 provinces (including municipalities and autonomous regions), and more than 1,000 counties and cities. China ranks the first in the world in terms of tea plantation and tea production. Tea plays a very important role in China's agricultural development.

1.5 The development of tea processing techniques and tea categories

Professor Chen Chuan from Anhui Agricultural University proposed a tea classification method, which has been widely accepted by academia and tea industry field. Based on Professor Chen's proposal, tea is classified into six major categories or groups, namely green tea, yellow tea, dark tea, black tea, oolong tea, and white tea, based on the processing technique and quality. When further processed with these six major groups of tea as raw material, such as scented tea, compressed tea, extracted tea and tea beverage, belong to the re-processed tea.

China has a long history of tea planting and making. China's well-developed traditional tea making techniques reflect the creativity of the Chinese nation. Tea leaves were initially picked to eat freshly as medical plant. During the process of using and making tea, Chinese people have accumulated experience to obtain a variety of tea making techniques. Over 2,000 tea varieties, mainly in six categories ——green, black, yellow, oolong, white and dark——are grown in China. Core skills include shaqing (enzyme inactivation), menhuang (heaping for yellowing), wodui (pile fermentating), weidiao (withering), zuoqing (leaves shaking and cooling), fajiao (oxidation or fermentation) and yinzhi (tea scenting). From the Western Zhou Dynasty to the Eastern Han Dynasty, and then from the Three Kingdoms Period to the early Tang Dynasty, the freshly picked tea leaves were sun-dried or pan-fried for dryness. During the Tang and Song dynasties, tea leaves were steamed for making Tuan Cha (ball-shaped tea) and loose tea. Tea leaves were steamed for enzyme inactivation, which is known as enzyme inactivation. Shaqing is one of the three processing techniques for making green tea. The others are Rounian (Rolling) and drying. Yellow tea was discovered coincidently when making green tea. After Shaqing process, people forgot rolling and drying tea leaves, resulting in yellow tea. Unlike green teas which undergoes minimal oxidation or black tea that is fully oxidized, yellow tea goes through a controlled partial oxidation process. The crafting of dark tea originated in the 16th century. Legend had it that when green tea in Yunnan and Sichuan Provinces was transported by caravan through The Ancient Tea-Horse Road to Xinjian and Xizang, the loose green tea was compressed in the shapes of cake and

and made a monumental impact to how tea was perceived and drunk — an impact that lasted for centuries to come. With his exceptional contribution in the development of Chinese tea, Lu Yu has been honored as the "Sage of Tea" by later generations.

1.3 The development of tea brewing and its preparing methods

Tea preparing methods have been constantly changing with the development of Chinese history. Each stage has its unique preparing method. The initial way of tea preparing is cooking with fresh tea leaves. During Tang Dynasty (618-907), the method by which tea was prepared is known as Jian Cha (boiled tea). During Song Dynasty, it was known as Dian Cha (whisked tea), and in Ming and Qing dynasties, it was known as Chongpao Cha (infused tea). The following will describe these preparing and brewing methods in detail.

During Tang Dynasty, the method by which tea was prepared is known as Jian Cha (boiled or cooked tea, in decoction). Once picked, the tender tea leaves would be pressed, and then poured into a mold to make into the shape of a cake for baking to dryness. From a cake of pressed tea leaves, a piece was taken, toasted and ground into powder, using a specific instrument for such use. Once ground, the powder was boiled. When it reached the desired degree of boiling, the concoction was served with a spoon into a bowl, and seasoned with salt, ginger and other ingredients, thus being ready to drink.

Different from the method of brewing tea during the Tang Period, in the Song Dynasty (960–1279), the prevalent way of having tea was through Dian Cha. The process begins with hot water being poured over fine powdered tea to create a paste, then more hot water is slowly added as the tea is constantly whisked by hand with a bamboo stick until froth appears. The unique whisking method of preparation along with tea-brewing competitions brought the art of tea to its pinnacle of refinement.

Before the Yuan Dynasty (1279 to 1368), people used to add a variety of ingredients when they were brewing tea. During the Ming Dynasty, Zhu Yuanzhang, the first emperor, gave an order to make a reform of "Gong Cha" System (the tea utilized as a tribute to the Emperor in Ancient China). This reform had relieved people's burden. Furthermore, Emperor Zhu encouraged people to produce a large quality of loose tea instead of compressed tea, which had facilitated tea production. New methods of processing tea (tea leaves being oxidized, steamed, and roasted etc.) gave a wide variety of different aromas, flavors, and even colors of the tea leaves and liquor. During the Ming and Qing dynasties, people brewed and drank tea by infusing processed tea leaves. People started to gain interest in the ever so varying tastes of tea and gradually abandoned ancient drinking methods. Infused tea is also the modern way of brewing and drinking tea.

1.4 The development of tea production

During the Jin and Southern and Northern dynasties, tea was produced in Sichu, Hubei, Hunan, Anhui, Jiangsu, Zhejiang, Guangdong, Yunnan and Guizhou provinces. During the reign of Emperor Yuan of the Jin Dynasty, local officials in Xuancheng County, Anhui Province has recorded that tea had been used as a tribute for the emperor. By the Tang Dynasty, the cultivation of tea trees had gradually spread from the inlands to the middle and lower reaches of the Yangtze River, which has became the center of tea planting and production at that time. Tea production areas have covered over 14 provinces in South Central China. In

In the *Section Sinensis* of *Theaceae* family, there are 31 species and four varieties in the world. China has 30 species and four varieties. China deserves the title of being the most abundant country in the tea tree germplasm resources.

1.2 The evolution of tea using and drinking

Tea was initially used as a medical plant before it becomes a popular beverage.

According to the record in *Shen Nong Compendium of Material Medic (Shen Nong Ben Cao Jing)*, one day Shen Nong was poisoned by 72 different poisonous plants, and found relief after tasting leaves from *Tu* (荼). *Tu* (荼) was the writing style of ancient Chinese character meaning *cha* (茶 , tea) in modern Chinese. Later on, tea herb has been constantly used by Chinese people as a medical plant for health benefits. In his work *The Tea List (Cha Pu)*, Gu Yuanqing, an expert of tea in Ming Dynasty (1368-1644), noted that "the benefits of drinking tea include quenching thirst, helping digestion, removing acne, refreshing mind, providing energy, supporting eyesight and relaxing body." Ethnic people living in the border area of China also use tea for health benefits. There is a popular saying by ethnic people, that is, "it's better to be deprived of food for three days than tea for one day." These ethnic people residing in high mountains and cold areas, such as Zang, Mongolian and Uygur, drink tea for digestion and nutrition, because their staple food are beef and mutton.

Tea leaves have been used as a kind of food by some ethnic people in China. Some ethnic people eat *Yancha* (Tea leaves being preserved and cooked by salt.), *Mingzhou* (Tea leaves used to cook porridge) and *Leicha* (Tea leaves used to be ground and mixed with sesame and peanut).

The history of drinking tea as a beverage is relatively not as long as that of using tea as a medical plant. The earliest record about tea drinking can be found in the book *Tongyue* written by Wang Bao, a verse and prose writer in Western Han Dynasty (206 B.C.–9 A.D.). He noted that "Buying tea in *Wuyang*, and brewing it with tea sets." The record showed that tea market and tea drinking were popular in Western Han Dynasty, which was more than 2,000 years. During over 500 years from Western Han Dynasty to Three Kingdoms Period (206 B.C. to 260 A.D.), only upper-class and noble-class families hade the privilege of using and drinking tea with the exception of people living in the Bashu area (today's Sichuan Province and nearby regions). Since Jin Dynasty (265-420 B. C.), tea came to be used and drunk by the middle and lower class family. From Jin Dynasty to the Southern and Northern Dynasties (from 265 A.D. to 581 A.D.), all sectors of society including rulers, officials, scholars, religious people and ordinary people generally started to drink tea, which showed that tea had become a custom and trend for Chinese people. Tang Dynasty (618 A.D. to 906 A.D.) was the golden age of tea drinking and using. Tea market became prosperous and tea drinking was popular in society. It is also during Tang Dynasty that the world's first monograph on tea emerged, that is *The Classic of Tea* written by Lu Yu. In *The Classic of Tea*, Lu Yu described the set of utensils for boiling and drinking powdered tea, thereby establishing the precedent for the importance of the wares and utensils used in tea ceremonies. Besides, Lu Yu recorded everything from how teas were grown and classified, to how teas should be chosen and cooked in *The Classic of Tea*, which became the first definitive guide to all tea-related thing. The book elevated tea to a new level, that is an art and culture,

Section Two The Origin and the Spread of Tea

It is believed that the writing style of the Chinese character 茶 (tea, pronounced as cha) emerged in the middle period of Tang Dynasty (618-907). 茶 (chá) TEA is one of the characters which conveys ideas and concept. There are nine strokes, and the upper part " 艹 " means "grass ", the middle part " 人 "means people, and the bottom part (木) means "wood". So TEA 茶 looks like a person picking tea leaves in the bush (between grass 艹 and 木). The tea plant is a woody plant, a member of the genus of *Camellia* in the family of Theaceae.

1. The Origin of Tea

1.1 The origin of tea plant

China is the home of tea plant. Study indicated that the origin of the tea plant could be traced back to southwestern China, including Yunnan Province, Guizhou Province and Sichuan Province, from which the tea spread around China and the world because of geological changes and cultivation. China is the first country to discover and consume tea, which is a major contribution to the world civilization.

Tea is indigenous to China. The discovering and using tea by Chinese can trace to Sheng Nong period which was more than 5,000 years ago. Legend has it that Shen Nong, the inventor of Chinese agricultural and Chinese medicine, tasted all kinds of plants in order to know the functions of herbs, and by coincidence, discovered tea herb. Research also showed that Chinese people began to cultivate tea plant as early as 3,000 year ago in the Western Zhou Dynasty. The consumption of tea in China has become popular since the 3rd. century A. D.

Known as the largest ancient tea tree in the world — one ancient tea tree located at Xiangzhuqing Village, Fengqin County, Lincang City, Yunnan Province, is as old as more than 3,200 years (See the picture). Its trunk is 1.84 meters in diameter with a height of 10.6 meters and waist of 5.82 meters. The aged tree, coupled with more than 14,000 ancient tea trees planted around it, is the living fossil that witnesses the long history of tea cultivation and tea drinking.

Ancient tea plant at Xiangzhuqing Village, Fengqing Country

"Tea can improve our physical and mental health and promote the harmonious development of society through its chemical and cultural elements." Tea is not only a substance with varied shapes, but a cultural element with profound meanings.

2.2 The combination of elegance and popularity

The development of tea culture is a combined process of elegance and popularity. The elegance of tea culture is obvious in these contexts, such as tea banquets held by high-class families, and tea tasting and appreciating conducted by literati and officialdom. These activities have resulted in poetry, song, dance, Chinese opera, calligraphy, painting, and sculpture. The popularity of tea culture is manifest at grassroots level, leading to the creation of folktales, legends, and sayings. However, elegance is based on popularity, which means that the former one will lose the basis for survival without the latter one.

2.3 The combination of functionality and aesthetic appreciation

Tea is extensively practical in satisfying human material life. For instance, tea is used for drinking, curing diseases and quenching people's thirst. Besides, tea is among the seven necessities for men of letters (music, chess, calligraphy, painting, poetry, wine and tea), and can satisfy people's spiritual need, which shows its extensively aesthetic features. Colorful tea flowers and various tea literary works, tea arts, tea ceremonies and tea etiquettes can also meet people's aesthetic needs. Therefore, tea, the unity of decoration, leisure and entertainment, not only displays art but also reflects folk custom.

2.4 The combination of practicability and entertainment

The practicality of tea culture determines its utilitarian nature. The comprehensive utilization of tea has developed and penetrated into many industries, for example, varied tea cultural activities including tea culture festivals, tea art performances and tea culture tourism, which have met people's need in tea tasting, leisure and tourism, and have boosted economy in recent years.

In summary, tea culture connotes a progressive view of history and world, encouraging human beings to realize their goals of social progress in a healthy, positive and peaceful way.

in China. Moreover, China has 56 ethnic groups, and each ethnic group has its unique way of practicing tea culture. Some ethnic people practice tea culture as a daily food and beverage for a balanced diet, such as tea with milk by Mongolian ethnic people, tea with milk and aroma by Uyghur ethnic people, and oiled tea by Miao and Dong ethnic people. Some ethnic people practice tea culture for personality development, such as the three-course tea by Bai ethnic people and the three-tea banquet by Miao ethnic people. Others practice tea culture for entertainment and enjoyment, for instance, the Bamboo-Tubed Tea by Dai ethnic people, the Thundering Tea by Lisu ethnic people, and the Clay-Pot Tea by Hui ethnic people. Still others practice tea culture for developing friendship, such as, the buttered tea by Zang ethnic people, the sour tea by De'ang ethnic people, and tea with milk by of Ewenki ethnic people.

1.4 Different tea culture practices in various regions and continents

As an ancient Chinese saying goes, "people in different areas share different life styles and customs." There are differences in the way of making, brewing and drinking tea. Generally speaking, Oriental people are fond of drinking pure tea without adding sugar, mint, lemon, milk, scallion, ginger or other condiments. People in European, American and Atlantic continents prefer black tea with milk and sugar, while people in West Africa and North Africa love green tea with mint or lemon. In China, southerners like to drink green tea, while northerners prefer scented tea. People in Fujian Province, Guangdong Province, and Taiwan Province enjoy Oolong tea. People in the southwestern part of China love drinking *Pu'er* tea. Ethnic people living in the border areas of China are fond of drinking beverages made of compressed brick-shaped tea.

1.5 The heritage of tea culture

Tea culture is an important part of traditional Chinese culture. With its social, extensive, ethnic and regional characteristics, tea culture as a heritage plays a role in promoting the development of Chinese culture. With the development of society, updated new culture elements and technology have been integrated in tea culture, which has enriched the connotation of tea culture. People love tea and tea culture because of its health benefits and rich culture connotation. Tea culture has become trendy more extensively and internationally.

2. The Connotations of Tea Culture

The history of tea culture demonstrates it has served to satisfy people's need materially and spiritually. Generally, there are four connotations of tea culture.

2.1 The combination of material and spirit

The saying goes, "Well fed, well bred", which implies that people will have good etiquette when they have met the needs of materials, such as food and clothing. It reveals that the enrichment of material and spiritual life will inevitably promote the development of culture. Wei Yingwu, a poet in Tang Dynasty (618 to 906 A.D.) wrote, "Its purity can not be defiled, and drinking tea can purify my soul." Su Dongpo, a famous poet in Song Dynasty (960-1279) said that "Good-quality tea is like a fair lady". Du Lei, also a poet in Song Dynasty wrote that "entertaining my guests in a chilly evening by drinking tea instead of wine." Lu Xun, a contemporary writer, believed that tasting tea is a kind of "pure happiness and luckiness". Japanese monk, Zen Master Eisai, said that

Section One Getting to Know Tea Culture

Tea culture is an umbrella term used to cover the material culture, the institutional culture, the behavioral and spiritual culture with tea as the carrier in the process of producing and drinking tea. From a broad perspective, tea culture covers both the science field and the humanities field of tea. From a narrow perspective, tea culture refers to the study of tea from the humanities, which focuses on the function of tea in cultivating human beings' spiritual and personal development in society.

1. Characteristics of Tea Culture

As a cultural phenomenon, tea culture covers both tea as the carrier of culture and various customs, values and beliefs of human beings while drinking tea. Thus, tea culture connotes both natural and social attributes. Its characteristics mainly involve the following five aspects.

1.1 Extensive tea culture practices in society

With the progress of society, tea culture has been practiced in all aspects of life. An ancient Chinese saying goes that "firewood, rice, oil, salt, sauce, vinegar and tea are the seven necessities of daily life". Chinese people also worship religious figures and ancestors with "three cups of tea and six cups of wine". These sayings and social rituals best illustrate the important role that tea culture played in society. Though having different tea drinking customs, people from various backgrounds and classes regard tea as a daily refreshment as well as a way of cultivating one's character and spirit development. In particular, celebrities, men of letters, and religious practice figures pursue seven practices in their life, namely music, chess, calligraphy, painting, poetry, wine and tea. Their pursuit in tea practice has promoted the development of tea culture.

1.2 Diverse tea culture practices in various countries

Tea culture has been practiced diversely in various countries in order to meet the different needs of people. The ancient Chinese tea culture, when combined with the local custom, has been further developed into the diverse tea practice in foreign countries. Currently, there are afternoon tea culture in Britain, Japanese tea culture, South Korean tea culture, Russian tea culture, Moroccan tea culture and so on. Tea culture practice can serve as a bridge connecting people from all over the world.

1.3 Colorful tea culture practices by ethnic people

According to a historical record, tea culture practice can find its root in Ba-Shu people living in Sichuan Province in ancient China. It was then mostly practiced by Han ethnic people

Chapter 1
Introduction to Tea Culture

Chapter 3
Health Benefits of *Pu'er* Tea // 031

Section One Lipid-Lowering and Weight-Losing Efficacies of *Pu'er* tea // 032
1. The Overview of Obesity // 032
2. The Efficacies of *Pu'er* Tea on Obesity // 033

Section Two Anti-Fatigue and Anti-Aging Efficacies of *Pu'er* Tea // 036
1. The Concept of Fatigue and Aging // 036
2. Anti-Fatigue and Anti-Aging Efficacies of *Pu'er* Tea // 036

Section Three Antioxidant Efficacies of *Pu'er* Tea // 039
1. Antioxidant Mechanism of *Pu'er* Tea // 039
2. Antioxidant Efficacies of *Pu'er* Tea // 040

Chapter 4
Drinking and Appreciating *Pu'er* Tea // 045

Section One Five Major Zones of *Pu'er* Tea in Yunnan Province // 046
1. *Pu'er* Tea Zone in Xishuangbanna District // 046
2. *Pu'er* Tea Zone in Pu'er District // 050
3. *Pu'er* Tea Zone in Lincang District // 051
4. *Pu'er* Tea Zone in Baoshan District // 053
5. *Pu'er* Tea Zone in Wenshan District // 054

Section Two Storage of *Pu'er* Tea // 056
1. Temperature // 056
2. Humidity // 057
3. Requirements for Other Storage Conditions // 057

Section Three *Pu'er* Tea Brewing and Tasting // 058
1. *Pu'er* Tea Brewing Skills // 058
2. Key Elements of *Pu'er* Tea Tasting // 067
3. The Assessment of *Pu'er* Tea // 083

Section Four Famous Tea Tree Planting Mountains and *Pu'er* Tea Brands in Yunnan Province // 088
1. Famous Tea Tree Planting Mountains in Yunnan Province // 088
2. Famous *Pu'er* Tea Brands in Yunnan Province // 100

Contents

Chapter 1
Introduction to Tea Culture // 001

Section One Getting to Know Tea Culture // 002
1. Characteristics of Tea Culture // 002
2. The Connotations of Tea Culture // 003

Section Two The Origin and the Spread of Tea // 005
1. The Origin of Tea // 005
2. The Spread of Chinese Tea Abroad // 009

Section Three The History of *Pu'er* Tea // 011
1. The Origin of the Name of *Pu'er* Tea // 011
2. The Historical Changes of *Pu'er* Tea // 011
3. The Historical Development of *Pu'er* tea Processing Techniques // 012
4. The Origin of Yunnan *Qizi* Cake-Shaped Tea // 013

Section Four Tea Custom in Various Countries // 015
1. Japanese Tea Ceremony // 015
2. South Korean Tea Ceremony // 016
3. Afternoon Tea in Britain // 016
4. Tea Drinking in Russia // 018

Chapter 2
Basic Knowledge of *Pu'er* Tea // 019

Section One Classification of Tea // 020
1. Comprehensive Classification of Tea // 020
2. The Three-Sphere Integrated Classification // 022

Section Two Basic Knowledge of *Pu'er* Tea // 024
1. The Concept of "*Pu'er* Tea" // 024
2. The Characteristics of *Pu'er* Tea // 024
3. The Classification of *Pu'er* Tea // 025
4. The Differences between Pure *Pu'er* Tea and Blended *Pu'er* Tea // 027

Preface

In order to introduce tea culture, especially basic knowledge and health benefits of *Pu'er* tea to domestic and foreign consumers, the team led by Professor Wang Baijuan from Yunnan Agricultural University (YAU) and Professor Wang Yuefei from Zhejiang University revised and updated the contents and data of the 2015 Edition *Drinking and Appreciating of Yunnan Pu'er Tea*. Funded by the Innovative Team and the Industry Innovator of *Xingdian* Talent Support Project in Yunnan Province, and translated by Professor Huang Yanhong, Associate Professor Wu Guangping from YAU, and Ms. An Shanshan from Kunming Shaoshan Elementary School, the Chinese-English version was published, hoping to further promote *Pu'er* tea knowledge and culture in Yunnan Province to tea lovers around the world.

This book is not only a book on the transformation of scientific and technological achievements, but also a popular science book in both Chinese and English for the global promotion of *Pu'er* tea and tea culture with Yunnan characteristics. It covers the following aspects:

Firstly, the book gives a brief introduction to the five current *Pu'er* tea growing and producing zones in Yunnan Province.

Secondly, the book provides a detailed guidance for consumers and tea lovers on how to choose high-quality *Pu'er* tea. Criteria and assessment are provided.

Thirdly, the book offers an elaboration of 16 mountains growing high-quality large-leaved tea in Yunan Province.

Finally, the details of *Pu'er* tea brewing and tasting are introduced. Several well-recognized *Pu'er* tea brands are introduced to help consumers to immerse themselves in *Pu'er* tea tasting in terms of color, infusion, aroma, after-taste, and infused tea leaves.

We would like to show our appreciation for all the members who have devoted themselves to reading, collecting and revising the data for the publication of the book. We are grateful for Professor Wang Yuefei from Zhejiang University. We also appreciate these tea industry companies for providing tea brewing and tasting samples. They are Yunnan Zhongcha Tea Industry Co., Ltd., Yunnan Six-Tea-Mountain Industry Co., Ltd., Wenshan Longlichun Tea Industry Co., Ltd., Fengqing Xiashan Tea Industry Co., Ltd., and Yunnan Defeng Tea Industry Co., Ltd. Our special thanks go to Mr. Zhang Jinggui, manager of Yunnan Fengning Tea Industry Co., Ltd., for his 5-year tea mountain field trip covering over 200,000 kilometers. He has personally paid many visits to tea mountains in Yunnan Province, taking pictures and collecting detailed information about these tea mountains, which have enriched the content of the book.

It is our hope that this book will help more tea lovers to get to know *Pu'er* tea, drink and appreciate *Pu'er* tea. *Pu'er* tea, as a specialty in Yunnan Province, will go from local to global.

Authors/Editors

Foreword

China was the first country to discover tea and draw benefits from tea, which in the long run, has been spread to various countries around the world. Nowadays, tea, coffee, and cocoa have been listed as the world's three major non-alcoholic beverages.

Globally, Yunnan is the initial center of tea trees. The Xiangzhuqing ancient tea tree in Fengqing County, Lincang City, is about 3,200 years old and is best known as the ancestor of tea on the earth. *Pu'er* tea is widely considered as original in Yunnan, where unique largeleaved tea trees are nurtured in an advantaged geographical and climatic environment. Coupled with distinctive processing techniques, locals have managed to make *Pu'er* tea a symbolic product moving towards worldwide. It has always been renowned both domestically and internationally for its long history, special quality and significant health benefits.

In the long development of tea in Yunnan, *Pu'er* tea has been constantly undergoing the baptism of time, changing its style, displaying its glamourous charm, and making more and more tea drinkers fall in love with it. This book briefly introduces the tea culture, the basic knowledge of *Pu'er* tea, the health benefits of drinking *Pu'er* tea and mainly focuses on the drinking and appreciating of *Pu'er* tea. It is a popular science book that provides guidance for readers in terms of drinking *Pu'er* tea properly. The main author of this book, Wang Baijuan, is one of my students, who has been enjoying great fame as an excellent Chinese tea connoisseur and a talent in the "Xingdian Industry Sector" Project in Yunnan Province. Moreover, she has unique insights into the taste of *Pu'er* tea. On the basis of years of practice of drinking multiple kinds of tea, she has consulted a large amount of relevant information, conducted a great deal of research on several large tea markets in Yunnan, visited tea merchants of major brands, and set her foot in plenty of best-known tea mountains before making this book in pubic.

This book is celebrated with its fluency in matters of writing and is easy to understand, which I believe, will lead more and more tea lovers to comprehend and delve into the world of *Pu'er* tea.

Wang Yuefei
Zhejiang University

Wang Yuefei

Wang Yuefei is a professor and doctoral supervisor in tea science at Zhejiang University (ZJU), Hanzhou, Zhejiang Province. He is now director of Tea Science Institute, ZJU, and a standing director of China Tea Association. Professor Wang has participated in a number of research projects, and has published more than 80 papers. He is the author, editor and co-editor of about 10 books. His research interests are the biochemical development of tea leaves, the health benefits and mechanism of tea, and the comprehensive use of tea resource.

Introduction to the Editors

Wang Baijuan is the professor and doctoral supervisor in tea science at Yunnan Agricultural University (YNAU), Kunming, Yunnan Province. She now is dean of College of Tea Science, YNAU, and a standing director of China Tea Association. She has participated in about 40 research projects and has published more than 100 papers. She is the author, editor or co-editor of about ten books, and many articles dealing with tea cultivation, processing, and tea culture. Her current research area focuses on smart tea cultivation and industry, and the application of big data on tea sector, which has achieved remarkable results by applying interdisciplinary and multi-perspective methods.

Wang Baijuan

Editors and Contributors

· Chief Editors ·
Wang Baijuan Wang Yuefei

· Co-editors ·
Zhang Guijing Sheng Yubo Deng Xiujuan Liu Chunyan

· Contributors ·
Shao Wanfang, Liu Xiaohui, Lü Caiyou, Zhang Zhigang, Ruan Dianrong, Wu Ya, Wang Xinghua, Li Jiahua, Hou Yan, Wang Jiayin, Chen Yaping, Li Jun, Zhao Shengnan, Yuan Wenxia, Chen Lijiao, Zhou Jianyun, Gao Jun, Jiang Bingbing, Zhao Chunyu, Huang Wei, Wen Bin, Wen Yan, Shen Xiaojing, Wei Zhenzhen, Qiu Mingzhong, Sai Shenli, Li Yanqin, Wu Tianyu

· Translators ·
Huang Yanhong Wu Guangping An Shanshan

Funded by
The Innovative Team of *Xingdian* Talent Support Project in Yunnan Province
The Industry Innovator of *Xingdian* Talent Support Project in Yunnan Province
The Innovative Team of Integrated Tea Culture, Industry & Technology for Rural Revitalization in Yunnan Province

Drinking and Appreciating Pu'er Tea

· Chinese-English ·

Edited by Wang Baijuan Wang Yuefei

Translated by Huang Yanhong Wu Guangping An Shanshan

Yunnan Science and Technology Press

·Kunming·